American Enterprise Institute

Automation Technology and Industrial Renewal

Adjustment Dynamics in the U.S. Metalworking Sector

Donald A. Hicks

American Enterprise Institute for Public Policy Research
Washington, D.C.

Donald A. Hicks is associate professor of sociology and political economy at the School of Social Sciences, the University of Texas at Dallas.

Library of Congress Cataloging-in-Publication Data

Hicks, Donald A.
  Automation technology and industrial renewal.

  (AEI studies ; 440)
  Bibliography: p.
  1. Metal-work—United States—Automation. 2. Metal-work—United States—Technological innovations.
  I. Title. II. Series.
  HD9506.U62H53   1986   671   85–31568

  ISBN 0-8447-3598-1
  ISBN 0-8447-3599-X (pbk.)

1 3 5 7 9 10 8 6 4 2

AEI Studies 440

Printed in the United States of America

Contents

FOREWORD *William J. Baroody, Jr.*   vii

PREFACE   ix

1 ADVANCED TECHNOLOGY AND MODELS OF INDUSTRIAL CHANGE   1

Interindustry and Intraindustry Adjustment   3
Manufacturing Automation: Implications for Work and
    Workers   5

2 MANUFACTURING AUTOMATION AND INDUSTRIAL CHANGE   10

Numerical Control Technology and Productivity   13
Automation Technology Diffusion in the Small Business
    Sector   15

3 THE U.S. METALWORKING SECTOR: AN OVERVIEW   27

The U.S. Metalworking Sector: A Setting for
    Change   28
The High-Technology Metalworking Sector   29
Establishment and Employment Change: 1980–1982   34
Small Metalworking Business Trends   41

4 VARIETIES OF INDUSTRIAL ADJUSTMENT   52

Industrial Adjustment: Plant Adaptation and
    Turnover   53
The U.S. Metalworking Sector: Background
    Characteristics   54

5 THE ADOPTION AND DIFFUSION OF AUTOMATED PRODUCTION
    TECHNOLOGY   81

Recent Evidence of Technology Diffusion   82
Cohort Patterns of Technology Adoption   84
To the Plant versus through the Plant:
    Diffusion versus Penetration and Saturation   89

Factors Influencing the Adoption of Automated Machine
   Control Technology 94
Technology Upgrading: CNC Programming Expansion
   Plans 99
Access Factors Governing the Adoption of
   Technology 102

6 AUTOMATION, EMPLOYMENT, AND WORKERS' ADJUSTMENT   108

Worker and Machine: Conflict and Coexistence 108
Patterns of Metalworking Employment Change,
   1980–1982 111
The Influence of Labor Environment 114
The Degree of Employment Change 116
A Closer Look at Employment Change 117
Responding to Employment Change 121
Accommodating Manufacturing Automation inside
   Plants 122
Conclusion 128

7 MAJOR FINDINGS AND POLICY IMPLICATIONS   132

The Limits of Technological Transfer and Capital
   Investment 133
New Tests of Corporate Rationality 136
Reappreciating Aggregate Industrial Adjustments 138
Technology Transfer and Industrial Policy
   Reconsidered 140
Employment Effects and Workers' Adjustment 141
Retargeting Policy Responses to Industrial Change 143

APPENDIX: Research Design   149

BIBLIOGRAPHY   153

LIST OF TABLES

1. Inventory of Productivity and Savings Opportunities
   Associated with NC/CNC Technology 16
2. High-Technology Employment Trends,
   1972–1982 30
3. Components of Establishment Change,
   1980–1982 36
4. Employment Gained and Lost by Formation, Failure,
   Relocation, and Contraction of Private Metalworking
   Establishments in the United States, 1980–1982 38

5. Small Business Share of the U.S. Metalworking Sector, 1980–1982   42

6. Employment Changes and Shifts in Employment Share by Establishment Size, U.S. Metalworking Sector, 1980–1982   45

7. State and Regional Distribution, Plant Universe and Sample, 1982   56

8. Date When Production Began at Plant, 1800–1982   59

9. Date When Production Began at Plant, by Region, 1800–1982   63

10. Corporate Structure in the Metalworking Sector, 1982   66

11. Corporate Structure in the Metalworking Sector, by Region, 1982   67

12. Date When Production Began at Plant, by Corporate Structure, 1800–1982   68

13. Region of Plant Component of Multiplant Company, 1982   69

14. Branching in the U.S. Metalworking Sector, 1982   70

15. Union Status of Production Workers, by Region, 1982   73

16. Union Status of Production Workers, by Right-to-Work Status of States, 1982   75

17. Date When Production Began at Plant, by Right-to-Work Status of States, 1800–1982   76

18. Date When Production Began at Plant, by Union Status of Production Workers, 1800–1982   77

19. Union Status of Production Workers, by Plant Type, 1982   78

20. Date of Adoption of NC/CNC Technology, by Year Plant Began, 1800–1982   86

21. Plants with Varying Numbers of NC/CNC Machine Tools, 1982   92

22. Technology Penetration: Machines in Plant under Alternative Types of Operation Control, 1982   93

23. Temporal Patterns of NC/CNC Diffusion among Types of Metalworking Operations, before 1950 to 1982   95

24. Rankings of Factors Prompting NC/CNC Adoption   98

25. Plant Plans to Expand CNC Programming Capabilities, by Region, 1982   101

26. Plans to Expand CNC Programming Capabilities, by Plant Type, 1982   102

27. Plans to Expand NC/CNC Machine Control, by Age of Plant, 1982   103

28. Plans to Expand NC/CNC Machine Control, by Union Status of Production Workers, 1982   103

29. Expected Length of Training Period   104

30. Expected Price of Computer-assisted Programming System   105

31. Plant Employment, 1980 and 1982   112

32. Employment Change by Age of Plant, 1980–1982   113

33. Employment Change by Region, 1980–1982   115

34. Employment Change by Union Status of Production Workers, 1980–1982   116

35. Employment Change by Right-to-Work Status of States, 1980–1982   116

36. Employment Change by Union Status of Production Workers and Plant Type, 1980–1982   118

37. Production Employment Change by Plans to Expand NC/CNC Machine Control, 1980–1982   122

38. Alternatives for NC/CNC Workpiece Program Preparation   124

39. Person Most Often Responsible for NC/CNC Workpiece Program Preparation   125

40. Type of Computer Support in Use   126

41. Average Weekly NC/CNC Workpiece Program Work Load   127

42. Average Number of People Preparing NC/CNC Workpiece Programs at Any One Time   127

43. NC/CNC Manufactured Parts Difficult to Program   128

Foreword

The challenge to the competitive standing of the United States in the international economy in recent years has prompted a reevaluation of the complex factors contributing to the competitiveness of American goods and services. Much research has focused specifically on the high-technology industries since they provide the cutting edge for U.S. international competitiveness. Few recent studies, however, have examined the process of technological change in the basic industries or so-called smokestack sectors of the U.S. economy. Yet the adoption and routine use of advanced manufacturing technologies in those sectors is crucial for U.S. industrial adjustment to changing domestic and international economic conditions.

This volume examines patterns of technological change and upgrading in one of the oldest industry complexes in the United States—metalworking—and compares them with alternative, nontechnological adjustments by American metalworking firms. The study focuses specifically on the adoption of automated process technologies within the sector, analyzing the many factors influencing the diffusion of technology as well as the effects on employment of technology upgrading.

Dr. Hicks challenges the common view that competitive pressures and technological change are destroying American basic manufacturing industries. He stresses that most new technologies seek to correct inefficient production arrangements, not to eliminate whole industries. Indeed, new technologies often facilitate industrial renewal in the basic manufacturing sector. The perception that equates new technologies and emerging high-technology industries with the destruction of the smokestack sector is simply wrong. The two are mutually reinforcing, not mutually exclusive.

Government policies designed to speed up the transition from basic to high-technology industries are therefore often misdirected. Governments, particularly state and local governments, should direct their attention to prospects for industrial renewal in the more traditional industries. More significantly, they should be aware of the important limitations to what we can expect from policies that promote technology transfer and capital investment in the hopes of creating a politically acceptable industry composition and historically familiar employment patterns. Policies promoting economic growth and new business formation may be more important for industrial adjustment than well-intentioned industrial policies dedicated to easing the adjustments of individual firms.

Donald Hicks's monograph is one of a series of conferences, seminars, publications, and special events sponsored by AEI's multiyear research project Competing in a Changing World Economy. The project is designed to examine structural changes in the world economy and to explore strategies for dealing with new economic, political, and strategic realities facing the United States.

WILLIAM J. BAROODY, JR.
President
American Enterprise Institute

Preface

According to the Bureau of Labor Statistics, 2.7 million jobs were lost in July 1981. In testimony before the Joint Economic Committee of the U.S. Congress, bureau Commissioner Janet Norwood illustrated the breadth of this decline by noting that nearly half the total job loss had occurred in machinery (standard industrial classification—SIC—35), primary (SIC 33) and fabricated (SIC 34) materials, and transportation equipment (SIC 37).[1] What had appeared two years earlier as an automobile industry recession eventually spread to a broad variety of manufacturing and service sectors anchored to locations throughout the nation, both within and beyond the traditional industrial heartland stretching from the Northeast through the Great Lakes region in the upper Midwest. So severe was the resulting employment contraction radiating out of the durable goods sector that between mid-1981 and the end of 1982 the United States experienced a *net* loss of 2.5 million jobs. Moreover, manufacturing employment shrank from 20.2 million to 18.2 million jobs while its share of total employment contracted from 22.1 percent to 20.5 percent.[2]

The linked sectors that constitute what is referred to here as the U.S. metalworking sector are the heart of durable goods production. These industries bore the brunt of the early and deep contractions that accompanied the arrival of the recessions of the early 1980s. For the most part, however, their arrival merely intensified and accelerated trends already at work throughout metalworking. The stage was set for witnessing the extent to which metalworking industries were capable of sustaining these effects—both cyclical and longer-term structural—and bouncing back. What on the surface looked like a classic case of deindustrialization and decline brought on by a restructuring domestic economy and intensified foreign competition was revealed on closer examination as a continuing test of the resiliency of an important segment of the nation's industrial base in adjusting to changing economic circumstances.

A long-awaited and broad-based recovery has now entered its fourth year. The industrial adjustment of the U.S. economy now under way is commonly assumed to be driven, at least in part, by the diffusion and adoption of new process technologies.[3] But the recovery, like the recession that preceded it, has been uneven in many respects. As a consequence, from a perspective that highlights technological change, many see the industrial landscape as being increasingly divided into "hi-tech" and "low-tech" production arrangements, the nation's economic and geographic landscape segmented into prospering and declining regional economies, and at least two generations of workers segregated by whether or not their skills are appropriate to the range of tasks that will dominate and define a newly configured economy. Most disturbing of all is the fear that these factors will be highly correlated, so that older industries, areas, and workers will be locked in a mutually destructive embrace while industries and workers in places both inside and beyond older industrial regions grow and prosper. Consequently, the causes and consequences of the development, diffusion, and upgrading of technology in industrial production settings are the central focus of this study.

In the metalworking sector, as elsewhere, any evidence of recovery has been accompanied by new pressures on individual metalworking plants and machine shops to increase productivity and to reposition themselves to face continued intense pressure from both domestic and foreign competition. As a result, an industry group like metalworking, which includes tens of thousands of small establishments, appears ripe for significant technology-led rationalization and restructuring. Adopting and incorporating new process technologies throughout individual industries is widely expected to be an effective strategy for many plants and shops as they continue to adjust to an economic climate that for them has long been hostile. Yet the rewiring of production technologies cannot be considered a panacea; even if it were, it might still not come about substantially. This study explores that prospect and finds that the relations between technological change and the continued industrial evolution of the United States can be complex and tenuous indeed.

Unavoidably, technology transfers and infusions into the production of goods and the provision of services are often both welcomed and feared. The trade-offs between enhanced productivity and international competitiveness and employment opportunity and security appear destined to accompany our exit from an earlier industrial era just as surely as they will accompany our passage through an advanced industrial one.[4] Although scientific advancements and technological innovations will surely figure prominently in the growth and development of new industries and new places in the future, the adaptation of advanced design, production, and related manufacturing technologies to older industrial arrangements likewise promises to contribute to the revitalization from within of the nation's existing industrial base.

Old industries will evolve into new industries. Revitalized older industries—and perhaps the regions in which they are located—can be expected to take their places alongside wholly new industries and regional economies in defining a new and advanced industrial era. Older physical capital will be not only jettisoned and discarded but also reorganized and thereby renewed. Older forms of labor relations, managerial approaches, and even definitions of "work" itself, however, will probably not make the transition. While this is the prospect that the present study offers, new production technology is best viewed as simply one expression—and maybe not the most important—of the kind of healthy industrial adjustment we are witnessing.

Organization of This Study

This study is guided by a model of industrial change that assigns primacy to the adjustments that unfold or are induced from within an industry and the individual firms and plants that compose it rather than to waves of new industries displacing older ones in an industrial hierarchy. Chapter 1 compares alternative models of industrial change and the roles each reserves for the process of technological innovation and transfer. The chapter also raises questions about the effects of the diffusion and implementation of new technologies on work, workers, and work settings.

The U.S. metalworking sector is not a single unified industry but a highly diverse industrial complex whose member industries share the common task of producing with metal. This complex includes industries that cover the full range of technological sophistication of both their products and their production processes. Chapter 2 introduces the notion of manufacturing automation as manifest in a quarter-century lineage of numerically controlled machine tool control systems (NC/CNC-CAM): numerical control (NC); computerized numerical control (CNC); and computer-assisted manufacturing (CAM). Since the data base developed for this study focuses on medium-sized and small metalworking plants and shops, an overview of the process of technological innovation in the small business sector is offered.

Chapter 3 presents a descriptive examination of recent enterprise, establishment, and employment trends in the U.S. metalworking sector. Its purpose is to explain what it is about the sector that has made it a potential target of technology upgrading in recent years. Chapter 4 explores various aggregate-level industrial adjustments that have taken place in the sector throughout the twentieth century. It gives special attention to the way in which the larger sector has been able to upgrade its physical capital plant continuously and move steadily from one labor environment to another—all without abandoning its historical regional concentrations. Ultimately, these larger-scale industrial adjustments are viewed as important contextual features for the process of adjustment to the adoption and implementation of automated and programmable production technologies.

The patterns by which NC/CNC-CAM technology has spread through the U.S. metalworking sector are presented and interpreted in chapter 5. An effort is made to distinguish various separate, distinct steps in the larger process of technology transfer. The influences of such factors as unionization, corporate structure, and regional location on the diffusion and implementation of automated production technology are also explored.

In chapter 6 automation is viewed as a cause rather than a consequence in the larger process of industrial change. The effects of the introduction of new technologies in metalworking plants and shops on workers, workplaces, and the notion of work itself are discussed. As background for this perspective, the chapter takes a closer look at the extent and structure of employment change experienced in medium-sized and small metalworking plants and shops during the 1980–1982 period.

Finally, chapter 7 attempts to summarize the principal findings of the study and their implications for identifying and interpreting the complex processes of industrial change. Moreover, it tries to suggest how these implications can inform and guide national and regional policies aimed at influencing advanced industrial development and the full range of industrial adjustments that process entails.

This book is part of a larger study of industrial change and the roles played in it by technology and information-based industries that I have been working on for the past four years. All along the way I have enjoyed the support and encouragement of colleagues, students, and friends. I wish to extend my appreciation to all of them and special thanks to a few. I am happy to acknowledge the exceptional cooperation and assistance of John Zinchak and Larry Patrick at UCCEL (formerly the University Computing Company) in Dallas, Texas. Without their efforts it would not have been possible to have access to the excellent data base constructed for use in this study. In addition, I wish to express appreciation to John Rees (Syracuse University) and my colleague Ron Briggs for their insights and the opportunity to design portions of this study in such a way that it could build on their recent work on the role of technological innovation in regional development. I also owe a debt of gratitude to Claude Barfield of the American Enterprise Institute, who invited me to prepare this study; to Herbert Fusfeld, Science Policy Center, New York University; Ken Gettelman, editor of *Modern Machine Shop*, Cincinnati, Ohio; and Michael T. Kelley, deputy assistant secretary for basic industries, International Trade Commission, at the U.S. Department of Commerce, who offered helpful insights and suggestions while the study was in its early drafts.

To Janet Browning, Sandy Seaberry, Akbar Torbat, Pam Van Cleve, and Kathy Kushner, I extend special thanks for their diligence in assisting me in numerous ways in the survey research and data analysis tasks of the study. For their assistance in manuscript and table preparation, I thank Cynthia Keheley, Debbi Gabar, Florence Cohen, and Phil Berry. For the special incentives that friends provide often unwittingly in helping see a task through to its completion, I thank Ron Lippincott of the University of Baltimore. Finally, for the support and encouragement she always shows as research and writing projects consume workdays, evenings, weekends, and vacations indiscriminately, I wish to thank my wife and best friend, Tanya. The interpretations and views expressed in this book are my own and in no way reflect the official corporate position of the UCCEL Corporation, the Numerical Control Society, *Modern Machine Shop*, or the American Enterprise Institute for Public Policy Research.

Notes

1. Cited in A. Swardson, "U.S. Jobless Rate Hits 10.8% in November," *Dallas Morning News*, December 4, 1982.

2. Bureau of Labor Statistics, *Centennial: 1884–1984, Employment and Earnings* (Washington, D.C., 1984), p. 85.

3. The relative importance of new technology to industrial adjustment is, however, the subject of considerable debate. Mass and Senge illustrate this with their suggestion that "encouraging industries . . . does not require finding new technological breakthroughs so much as sustaining social and economic conditions that can make existing inventory economically attractive." N. J. Mass and P. M. Senge, "The Economic Long Wave: Implications for Industrial Recovery," *Economic Development Commentary* (National Council for Urban Economic Development) (Spring 1983), pp. 3–90.

4. The dominant features of advanced industrial development are explored in D. A. Hicks, *Advanced Industrial Development: Restructuring, Relocation, and Renewal* (Boston: Oelgeschlager, Gunn & Hain, forthcoming), chap. 1.

1

Advanced Technology and Models of Industrial Change

The development and diffusion of new technologies have come to be viewed in recent years as important, if not indispensable, ingredients in the adjustment of older industrial economies like that of the United States to new global economic realities. By the early 1980s the conventional depiction of modern industrial change involved a changing composition of leading industries in which many older mainstays— symbolized most dramatically by the automobile and steel industry complex—would gradually yield to a wave of new industries that harnessed the abilities of new clusters of frontier technologies, including lasers and fiber optics, genetic engineering, robotics, computer hardware and software, photovoltaics, and new materials processes such as ceramics and powdered metallurgy. From this perspective industrial change reflected a kind of mechanical circulation process, in which the relentless evolution of new technologies led to the gradual rotation of newer industries into the leadership roles in growth and productivity once held by older and now beleaguered "sunset" industries.

So complete was this fascination with a small cluster of "sunrise" industries derived from new technologies that the dynamics of industrial change were typically traced to interindustry, rather than intra-industry, shifts. Despite building evidence to the contrary, industrial change came to be widely viewed as located between industries rather than within them, as though any particular industry were destined to travel a fixed course dictated by its product cycles and production technology life stages to an inevitable diminution or demise.

In one sense this fixation on new technology—whether embodied in new production (that is, process) technologies or in the products made possible by them—signifies little more than the periodic redis-covery of the significance of tools, their mechanization, and their organization in the larger sweep of industrial change.[1] In another sense, however, economic developments during the 1970s and early 1980s offered an unusually accommodative backdrop against which to

1

appreciate the potential of new technologies. The stagnation of productivity growth, the declining employment share in manufacturing, the cessation of employment growth in many older manufacturing industries, the acceleration of new high-technology business formations, and the explosive commercialization of the microcomputer all combined to attract great attention to the industrial potential of new technological developments. Processes of technological innovation, which have always been part of industrial change, gained new visibility in large part because they continued during a period when they were no longer dominated by expansion in the far larger—and older—industrial sectors. A view of industrial change has thus arisen in which the primary dynamic is the rise of new industries closely tied to new technologies. Consequently, technological change has become the target of active promotion rather than simply passive acknowledgment.

The emergence of the view that a succession of new technologies is the sine qua non of advanced industrial development is not without justification. In a long and venerable tradition, technological change has been identified as the prime mover in the evolution of capitalist economies and industrial development. Although their work is separated by many years, Schumpeter and Rosenberg stand out as analysts who assigned primacy to technological change in processes of economic growth and industrial development. But a congressional report on the growth and development of the so-called high-technology industrial sector issued in mid–1982, just as the most recent recession began to wane, set off a new avalanche of interest in the influence of technological advances on economic growth as registered in expanding output, employment, and incomes.[2]

Among the findings of the congressional report was that of the net increase in U.S. manufacturing employment from 1955 to 1979, three-quarters was in high-technology industries. The implication was that a complex of older manufacturing industries—long regarded as the core of our industrial economy—had gradually fragmented and was yielding to a new collection of goods- and services-producing industries whose investment patterns and occupational profiles reflected their heavy dependence on basic scientific and applied research and development (R&D) performed by a highly skilled work force that could keep those industries at their technological frontiers. More recently, Lawrence has not only confirmed the existence of this structural shift in U.S. manufacturing but also demonstrated that it was well under way as early as the 1950s. High-technology industries alone had increased their output and employment shares during 1960–1980, a structural shift that came at the expense of all other

2

segments.[3] Finally, just as the thickening interdependencies between the United States and the rest of the world were being acknowledged, the decisive role of technological innovation in the global economy was clearly registered in the dramatic shift in the U.S. trade balance to R&D-intensive manufactured goods.[4]

Innovation and Industry Change. The new-found respect for science and engineering, research and development, and their derivative technologies has been limited largely to the innovation process. Although single-factor theories of industrial change should by now be suspect, waves of new industries tied to new technologies have captured widespread attention, as if the sheer momentum of the evolution of a technology were sufficient to launch an industry in its wake.[5] The view that these new industries and the structural shifts they define are better understood as economic than as purely technological phenomena has been largely obscured. Commercializing innovation ultimately depends on whether sufficient demand exists in older or newly formed markets for a new product.[6]

This perspective raises the possibility that an even more compelling view of the role of technology in industrial change involves the diffusion and implementation of new (and often not-so-new) technologies throughout older industries: "For it is not the spectacular innovations (crucial though they may be as the turning points) which are the important elements ... but the *diffusion* ... and the rate of such diffusion."[7] An alternative view of how larger-scale industrial change unfolds is then possible. The diffusion of new technologies can lead to the renewal of older industries, not just the birth of new ones. The focus of industrial change then shifts to what is happening within industries and the firms that constitute them and away from an exclusive preoccupation with the eclipse of entire older industries by newer ones. In the end the full sweep of industrial change is seen to involve not only the new growth tied to industrial innovation but the prospect of renewed growth tied to the industrial adjustment of older industries.

Interindustry and Intraindustry Adjustment

It is not new technologies per se, then, but new applications of technologies in ways not heretofore undertaken that interest us here. And it is new applications in their roles as catalysts of industrial rejuvenation and renewal that are the principal focus of this study. As we shall come to see, however, industrial adjustment is a layered process. The adjustment of an older industry—like that of an entire economy—can proceed simultaneously on two distinct levels. The first is what happens between firms. Through the continuous replacement of older, uncompetitive firms by newer, more competitive ones and the unequal expansion and contraction among existing ones, the composition of an industry can change dramatically over time. As a result, industrial adjustment unfolds in the direction of business formations and expansions and away from business failures and contractions. Such replacement dynamics are known to speed up during recessions and longer periods of structural change as older industry forms are unable to compete in changing economic circumstances. This is the stuff of "shakeouts," about which so much is heard these days. It is illustrated by the structural change in the steel industry resulting from the shift of growth to minimills producing specialty products and away from large, integrated steel mills.

The second level of industrial adjustment involves what happens within firms. Here such things as decisions to adopt new technologies and new management tools with which to implement them play a central role. At issue is whether the adjustment of the individual firm will enable it to survive in a changing economic climate. The reconceptualization of the century-old process of automobile production as symbolized in the General Motors Saturn Project illustrates this second kind of adjustment. Ultimately these two levels, though distinct, are linked to each other. That is, the adoption of new technologies by firms in older industries and the degree to which they are diffused across the full range of production tasks in individual firms do not take place in isolation. A major message of this study is that intrafirm changes, such as the adoption and implementation of automated production technologies, must be viewed against the backdrop of interfirm adjustments.

The evidence is clear that the diffusion of automated manufacturing technologies in the form of NC/CNC-CAM (a family of numerically controlled machine tool control systems) has been only one way—and perhaps not the most important—in which the U.S. metalworking industry complex has changed in recent decades. Major unplanned and unorchestrated adjustments, such as the filtering of production activities in the larger sector into ever newer plants and shops and the filtering out of older labor relations arrangements, have accompanied the more consciously engineered strategies of capital investment, including the adoption of ever newer forms of factory automation. Moreover, these varieties of capital mobility have been expressed in organizational, rather than locational, changes. The industrial adjustments in the metalworking sector examined in this study have not led to a lessening of the concentration of the industry

4

in the East North Central and Middle Atlantic regions, a fact that has important implications for how the complexities of industrial adjustment are interpreted economically and socially and responded to politically throughout the larger society.

Manufacturing Automation: Implications for Work and Workers

Intrafirm adjustments such as the upgrading of production technologies can be viewed as results of a complex array of explicit and subtle incentives. Certainly the investment tax credit, capital gains taxes, corporate income tax, and accelerated depreciation allowances in the tax code can influence the course and pace of such adjustments. In addition, special cost pressures tied to shortages of certain labor skills or the presumed inflexibility of organized labor may influence whether and to what extent manufacturing automation takes place. Thus the importance of automation as a consequence of certain aspects of industrial change sets the stage for consideration of automation as a cause of yet other aspects of industrial change involving not only the workplace but also the worker and the very definition of "work."

Work and Workers in the Automated Work Setting. Early forms of industrial development had the notable effect of concentrating both production and its administration in common locations. The factory—like the industrial-era city that housed it—marked a coming together of productive arrangements that in the beginning appeared as contrived and artificial as it was rational. Once those arrangements were gathered in a common location, the principal organizational effect of mechanized mass production was first to fragment and multiply the number of work tasks and only later to organize them into a complex division of labor. In contrast, contemporary advanced industrial development has been facilitated by enormous technological capacities that have permitted the deconcentration of standardized production activities at several spatial scales amid continuing administrative centralization of managerial control.[8] Yet an equally important effect of new production technologies inside plants has been to shrink the number of separate tasks and simplify the resulting industrial division of labor.[9] Precisely this kind of complexity of effects on the substance of work holds great significance for workers and work settings.

The redefinition of work and workers' skills. One of the major implications of factory automation—like automation on the farm in the past or in the office more recently—is that work activities and the settings

5

in which production work takes place cannot escape redefinition, reorganization, and even relocation. Early in the urban-industrial era, work and the role of the worker were increasingly characterized by the skills required by different tasks. The traditional role of craftsman survived even while the location of work moved out of the home and into a segregated workplace devoted to small-lot batch and later mass production. In manufacturing, even though tasks were gradually fragmented, at many stages of the production process the precision and care reflecting the qualities of experience and craftsmanship were retained and remained invaluable.

The slowly changing metalworking sector is a useful setting in which to examine the redefinition of work and the shifting demand for specific skills. Even the arrival of the assembly line and standardized parts did not dilute the importance of the skilled craftsman until well into the twentieth century. World War II stands as the watershed of that evolution. During and immediately after the war the design of instruments of national defense and of waging war itself began to incorporate increasingly advanced technological features. The shift to reliance on aircraft and new kinds of naval vessels and armored vehicles, as well as shipboard wizardry and armaments for all of them, illustrates this well.[10] The engineering and scientific contributions to military and domestic manufacturing designs increased substantially. Greater quality control, much reduced tolerances, and greater need for precision over long and short production runs placed complex new pressures on traditional manufacturing arrangements and production technology.

In the metalworking sector this pressure led to a gradual shift from conventional or manual control of machine tools to new forms of nonmanual or automated control, such as NC/CNC-CAM, in selected industries. The embryonic computer was harnessed in the early 1950s in an effort to reduce the variation in the final product introduced by the human factor through the "craft" of machine tool control.

> Traditional manual machine operation is largely based on intuition and skills. These skills are learned through long training and practice and, at their highest level, they are unsurpassed in turning out fine work. The final product is, however, heavily dependent on the machine operator's skill and knowledge, and these qualities are never fully consistent.[11]

With this development came the inescapable redefinition of skills and work in many of the nation's basic industrial sectors—durable goods manufacturing in general and metalworking in particular.

6

Automation: cause or consequence? As happened when labor was gradually wrung out of a rapidly rationalized agrarian economy, shifting demands for specific kinds of labor in goods production are finally beginning to reflect the new truncated organizations of work. Unskilled labor, which was once the mainstay of the factory environment, is less and less in demand.[12] Once again we see evidence that clusters of workers defined by age, tenure, and location in declining industries in declining regions are being caught in an industrial maelstrom.[13]

Predictably, great debate surrounds questions related to whether NC/CNC-CAM and related forms of manufacturing automation follow or precede the declines in productivity and employment that eventually stalk obsolete physical plant and equipment, deferred capital investment, reduced R&D, and inappropriate industrial organization. Do plants faced with the option of rationalizing their shop floor production arrangements adopt advanced production technologies after their competitive position begins to slip and employment contracts, or do they upgrade their technology to reduce labor bills through labor substitution in their quest for increased productivity and the restored competitiveness it may offer?[14] These are complex questions, and the causal sequences in most data sets are difficult to sort out. For that reason, as I consider the influence of technology upgrading and industrial renewal in a later chapter, I give careful attention to the effects of automation on the world of work.

Notes

1. O. Mayr and R. C. Post, eds., *Yankee Enterprise: The Rise of the American System of Manufactures* (Washington, D.C.: Smithsonian Institution Press, 1981); and E. Ginzberg, "The Mechanization of Work," *Scientific American*, vol. 247, no. 3 (September 1982), pp. 39–47.

2. J. Schumpeter, *Capitalism, Socialism, and Democracy* (New York: Harper and Brothers, 1942); N. Rosenberg, *Technology and American Economic Growth* (New York: Harper and Row, 1972); and R. Premus, *Location of High Technology Firms and Regional Economic Development*, Staff study of Joint Economic Committee (Washington, D.C., 1982).

3. Not surprisingly, once again the lag between a development and its discovery is revealed in Lawrence's analysis. His data indicate that the average annual growth rates in high-technology employment, value added, and productivity during the 1960s actually exceeded those realized since then, indicating that the pace of restructuring was greater a quarter-century ago than it has been since. Further evidence of the restructuring of the U.S. economy is found in Lawrence's demonstration that the driving force behind the "declines" in the domestic automobile and steel industries during the early 1980s

was the restructuring of domestic demand for these products rather than the inroads made by foreign imports. See R. Z. Lawrence, *Can America Compete?* (Washington, D.C.: Brookings Institution, 1984).

4. Office of the U.S. Trade Representative, *Annual Report of the President of the United States on the Trade Agreements Program* (Washington, D.C., 1984), p. 16.

5. C. Freeman, J. Clark, and L. Soete, *Unemployment and Technical Innovation: A Study of Long Waves and Economic Development* (Westport, Conn.: Greenwood Press, 1982); and R. Gilpin, *Technology, Economic Growth, and International Competitiveness*, Report to the Joint Economic Committee (Washington, D.C., 1975). Freeman et al., in a discussion of "new technology systems," observe that "once the industrial application of a new technology begins to develop, to a considerable extent it has its own momentum" (p. 72). For a discussion of "technological trajectories," see R. R. Nelson and S. G. Winter, "In Search of a Useful Theory of Innovation," *Research Policy*, vol. 6 (1977), pp. 36–76.

6. For a brief discussion of the "demand-pull" and the "technology-push" models of technological development, see L. G. Tornatzky, J. D. Eveland, M. G. Boylan, W. A. Hetzner, E. C. Johnson, D. Roitman, and J. Schneider, *The Process of Technological Innovation: Reviewing the Literature* (Washington, D.C.: National Science Foundation, 1983), p. 182.

7. D. Bell, *The Coming of Post-Industrial Society: A Venture in Social Forecasting* (New York: Basic Books, 1973), p. 318 (italics added). Various targets and implications of industrial adjustments in the older automobile and steel industries are discussed in Office of Technology Assessment, *U.S. Industrial Competitiveness: A Comparison of Steel, Electronics, and Automobiles* (Washington, D.C., 1981), p. 69.

8. D. A. Hicks, *Advanced Industrial Development: Restructuring, Relocation, and Renewal* (Boston: Oelgeschlager, Gunn & Hain, forthcoming); and R. Cohen, "The Internationalization of Capital and U.S. Cities" (Ph.D. dissertation, New School for Social Research, 1979). See also the discussion of the rise of complexes of corporate activities and their locational features in T. J. Noyelle and T. M. Stanback, Jr., *The Economic Transformation of American Cities* (Totowa, N.J.: Rowman and Allanheld, 1984), chap. 6.

9. See related discussion in K. E. Gettelman, M. D. Albert, and W. Nordquist, "Introduction: Fundamentals of NC/CAM," in Gettelman et al., eds., *Modern Machine Shop: 1985 NC/CAM Guidebook* (Cincinnati: Modern Machine Shop, 1985), pp. 37ff., 176ff.

10. Accompanying the structural shift to more sophisticated weaponry was a spatial shift: by the 1950s increasing proportions of the defense budget were committed to prime contracts and military payrolls filtering out of the industrial heartland and into regions west and south. Even the tendency for these resource flows to course back into the Northeast and Midwest through elaborate subcontracting networks did not reduce the effects on the economy of increasingly sophisticated defense requirements. For an analysis of these trends, see J. Rees, "Government Policy and Industrial Location in the United States," in Joint Economic Committee, *State and Local Finance: Adjustments in a*

Changing Economy, Special Study on Economic Change (Washington, D.C., 1980), vol. 17.

11. K. M. Gettelman and M. D. Albert, "Introduction: Fundamentals of NC/CAM," in Gettelman and Albert, eds., *Modern Machine Shop: 1982 NC/ CAM Guidebook* (Cincinnati: Modern Machine Shop, 1982), p. 307.

12. For a discussion of this trend, see Office of Technology Assessment, *Computerized Manufacturing Automation: Employment, Education, and the Workplace* (Washington, D.C., 1984), chaps. 4–6.

13. See Congressional Budget Office, *Dislocated Workers: Issues and Federal Options* (Washington, D.C., 1982). See also R. Hanson, ed., *Rethinking Urban Policy: Urban Development in an Advanced Economy* (Washington, D.C.: National Academy Press, 1983), pp. 101ff.

14. The tension between these two perspectives is briefly illustrated in Joint Economic Committee, *Robotics and the Economy* (Washington, D.C., 1982).

2
Manufacturing Automation and Industrial Change

Today we hear much of automated factories, flexible manufacturing systems, and computerized design and machine tool control systems. Although each of these process technologies rightly commands attention, the relations among them and their implications for all facets of industrial production are arguably more important. The emergence and development of these individual technology forms have not been random or haphazard; indeed, they can be arrayed sensibly along one or more technological time lines covering the second half of this century. As figure 1 suggests, the lineage of advanced manufacturing technologies exhibits a developmental or life-cycle quality. In retrospect it is easy to imagine how the momentum of such a progression of increasingly sophisticated technologies might appear able to assert itself by sweeping through manufacturing industries and transforming them.

Yet the evolution of increasingly automated manufacturing technologies cannot be understood apart from the larger economic and social contexts of which they are a part, which have prodded them into existence and have drawn them into application. The slow pace with which U.S. factories in many industries have been automated illustrates their greater capacity to manipulate materials than information, as well as the disruptive effects on current production of implementing new and expensive technologies.[1] It is not enough that a new technology offers a better way of doing something. It is also important that technologies fit the requirements and capabilities of actual production arrangements inside workplaces. Where innovative technologies lack that crucial fit, they are not likely to be diffused easily into and through plants and shops throughout an industry.

Although a technology may evolve, its effects on the industrial arrangements or specific operations for which it is intended may be blunted by the structure of the target industry, its characteristic labor patterns, and its own scale, price, and performance characteristics. Moreover, as in the machine tool industry, which supplies factory

10

FIGURE 1

HISTORICAL DEVELOPMENT OF CAD/CAM SOFTWARE CONCEPTS

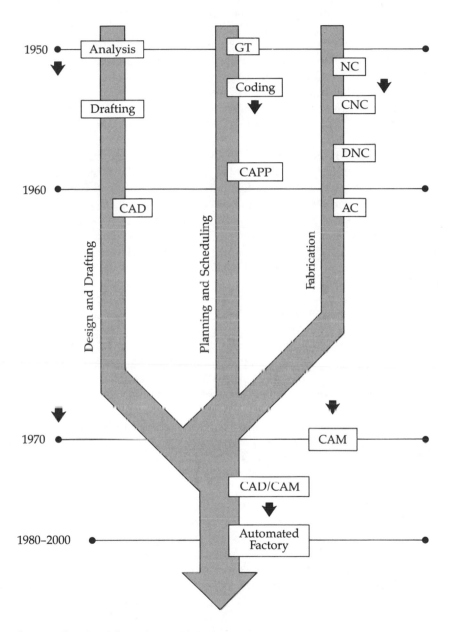

SOURCE: Reprinted from the *Machining Data Handbook*, 3d ed., by permission of the Machinability Data Center. ©1980 by Metcut Research Associates Inc.

11

automation equipment to the larger metalworking sector, the intense competition in regionally concentrated markets, the prevalence of small, traditional family-owned firms, and the difficulty of channeling capital into investment amid rapid and frequent cyclical shifts can dampen technology upgrading by manufacturing firms.[2]

To understand better how the pace and patterns of technological development are effectively channeled by the defining features of an industry and its economic environment, let us consider the innovation and diffusion of a family of numerically controlled machine tool technologies—numerical control (NC) and computerized numerical control (CNC)—throughout the U.S. metalworking sector.

New Waves of Manufacturing Automation. A recent study by the Office of Technology Assessment (OTA) reported that the manufacturing sector is poised to experience sweeping changes through the adoption of programmable automation capabilities such as are found in robots, computer-aided process planning (CAPP), computer-assisted design (CAD), and computer-assisted manufacturing (CAM).[3] Yet, although the possibility of harnessing computers to industrial production processes is not new, the diffusion of such upgraded technologies to and through factories has generally been limited to firms that are large enough to accommodate the typically heavy initial investments and oriented to volumes and combinations of manufacturing operations that make their automation economically feasible.[4] Over the years the adoption by individual firms of NC/CNC and related CAD/CAM systems has often required the investment of hundreds of thousands of dollars. Very recently, however, continued development of components of CAD/CAM systems has allowed manufacturers of these systems to compete through product differentiation and process changes as well as price.[5] The result has been a closer tailoring of specialized products for the targeted user as well as a predictable plummeting of prices. Accompanying the price declines has been a dramatic expansion of the potential markets for these systems.[6]

The continuing development of automated manufacturing technologies can be viewed as a response to the rise of new markets and new levels of demand in older markets. Today rising production costs faced by metalworking firms, chronic shortages of skilled machinists and related labor specialists on whom conventional machine control methods depend, stiff competition from foreign producers, and contractual stipulations that frequently require ever greater quality control have increased the incentives that can "pull" manufacturing automation technologies through stages of innovation and lead to their

12

diffusion and implementation through a wide variety of metalworking industries. New forms of existing technologies have thus begun to be increasingly better matched to the requirements of thousands of manufacturing establishments for which they were previously inappropriate. With the development of relatively inexpensive NC/CNC and CAD/CAM systems, many new forms of automation are now within the reach of medium-sized and small manufacturing and engineering firms.[7] Much advanced manufacturing technology will now probably filter down through industries in the larger metalworking sector after having been hindered in doing so for over a quarter-century. The ability of this economic perspective on the technology life cycle to help us understand the processes of technology diffusion and industrial evolution appears great indeed.

Numerical Control Technology and Productivity

Numerical control is a generic term for a succession of technologies that code planning and programming design and production instructions into numbers and other symbols, which then govern machine tool operations and related production tasks. Based on pioneer work done in the 1940s, the technology was developed intensively as a result of air force contract support of the MIT's Servomechanisms Laboratory in 1949 for the purpose of finding a way to machine multiaxis air foil curves (such as wings and propeller blades) with greater standard accuracies than could conventionally be achieved by skilled craftsmen controlling machine tools by hand. The first NC machine tool became operational in 1952, and since then the technology has been available for adoption throughout the broader sector.[8]

Computerized numerical control (CNC) is a more recent expression of NC. Harnessing a microprocessor or computer module to a numerical control system permits machines to be controlled through fixed or variable control programs stored in the computer's memory. As figure 1 indicates, CNC was developed in the late 1950s and was available commercially by the mid-1960s. Recent CNC adoptions have largely eclipsed those of older forms of numerical control, especially among plants seeking to upgrade their older NC capabilities. Even more recent life stages in the NC technology sequence reveal the expanded importance of computerized automation; they include direct numerical control (DNC), more broadly defined computer-aided manufacturing (CAM), and integrated computer-assisted design and manufacturing (CAD/CAM). As it is now envisioned, the sequence is expected to culminate in the fully automated factory by the twenty-first century. Prototypes already exist in Japan, and the synergy of

General Motors and Electronic Data Systems in the Saturn venture symbolizes the intentions of major U.S. manufacturers to move boldly in that direction.

NC/CNC-CAM Production Efficiencies. In 1981 the Numerical Control Society (NCS)—a national organization dedicated to the development and implementation of NC/CNC-CAM in modern industry—compared the efficiency and effectiveness of NC machines with those of conventional machines. Table 1 lists ways in which numerical control capabilities can be translated into gains in productivity and related savings of space, material, time, and money. "Efficiency" was defined as the ability to perform close to the engineering standard hour. Past comparisons of NC with conventional machining methods had shown that in a ten-hour period NC machines could produce eight standard hours while a manual or conventionally controlled machine could produce only six. The NC study found that on the average NC machines are 10 percent more efficient than machines under manual control. "Effectiveness" was defined as the ability of a machine to make a quality product more predictably, as well as to produce more parts per hour. In the NCS study the efficiency of NC machines consistently outweighed that of manually controlled machines.

The shift of advanced manufacturing away from mass production, with its long runs and standardized output, in the direction of smaller plants, shorter runs, and specialized and customized production may well function to undercut these advantages, however. A recent Commerce Department study concluded that "low volume production, making only one or a few parts, is still generally most cost effective when performed by a skilled machinist on a manually controlled machine tool."[9] Ironically, illustrating what Samuelson has called the "fiction of corporate rationality," over half the companies in the NCS study report that they made no effort to measure the performance of their NC or manually controlled machines.[10]

Distinguishing between Diffusion and Implementation. NC technology passed from the invention to the innovation stage over three decades ago. While it has been widely adopted across the range of individual metalworking industries, its effect on the bulk of the activities in most industries has been relatively small. The extent of adoption can therefore be a somewhat misleading criterion for measuring diffusion. A distinction needs to be made between the innovation and diffusion of NC technology, in which a plant controls a metalworking operation by some form of NC, and measurable aspects of the implementation of that technology.[11] Later in this study I use the term "penetration" for the degree to which an individual plant has adopted some form of manufacturing automation and the term "saturation" for the share of a plant's total workload that has been so automated.[12] Implementation, therefore, directs our attention beyond the number of plants and shops that have adopted forms of NC/CNC-CAM and more toward considerations of the number of metalworking operations converted from conventional to numerical control and the role of automation in the larger production process. Ultimately, the implications for the productivity of a particular plant, as well as for the industry at large, depend on how completely the work performed inside a plant has been automated. A plant with a single numerically controlled machine tool among many conventionally controlled tools can be expected to experience far less of an effect on productivity than a plant where the mixture tilts in the other direction. If a plant reserves the use of numerically controlled machines for special or infrequent operations, the productivity of the plant will likewise not be appreciably increased.

Extent of NC/CNC-CAM Diffusion. Although NC technology has been adopted by many plants and shops throughout the metalworking sector, researchers at the *American Machinist* have estimated that only 103,000 (less than 4 percent) of the 2.9 million machine tools (2.3 million metal-cutting and 0.6 million metal-forming) installed in the United States were numerically controlled as recently as 1983, a number that nonetheless had tripled between 1973 and 1983 and doubled since 1978.[13] Moreover, many of the adoptions took place many years ago. As the 1980s began, over a third (34 percent) were at least twenty years old, a higher average age than that for any other major industrialized nation. This suggests that the lag between innovation and diffusion has been an extended one. For some plants, particularly large ones in the aircraft and related defense industries, diffusion has been relatively rapid and widespread; for much of the rest of the metalworking industry complex, it has been much slower.[14] Furthermore, even where diffusion has occurred in many plants and shops in an industry, implementation has proceeded slowly at best. While this study explores the role of several factors in facilitating or hindering the diffusion and implementation of NC/CNC-CAM, it gives special emphasis to these processes among medium-sized and smaller metalworking plants and shops.

Automation Technology Diffusion in the Small Business Sector

The effect of technology on production and productivity has become a more complex question as it has simultaneously become a more important one. Technology "adoption" decisions at a microeconomic

TABLE 1
INVENTORY OF PRODUCTIVITY AND SAVINGS OPPORTUNITIES ASSOCIATED WITH NC/CNC TECHNOLOGY

	Anticipated Savings from NC Machine		*Anticipated Savings from NC Machine*
Improved accuracy	5% of direct labor cost	Control of cycle in hands of management—can be fixed	10% increased production
Reduced cutting tool adjustment by use of tool offsets	5% of direct labor cost	Savings in setting and maintaining standards	50% of cost of standards
Reduced cutting tool change time—change only when dull	20% of tool allowance	Power consumption more level because of continuous running	5% of power cost
Reduced cutting tool cost—throw away carbides—more standard tools—less specials	25% of tool cost	Reduction of inventory	5% of dollar value of inventory
Longer tool life due to optimum cutting speeds and feeds	30% of tool cost	Savings from storage of less productive material	20% of storage area
Savings in purchasing—less tools—less paper	5% of tool cost	Less inventory—less material handling	5% of material handling cost
Improved tool life due to improved machine performance	20% increased tool life	Floor space savings due to need for fewer machines	Actual space saved
Reduce cutting tool storage—simpler tooling	50% of tool crib area	Savings in supervision	Actual number saved
Savings in tool maintenance—cutter grinding	20% of cutter grinding cost	Lower fringe costs due to more productive time	25% reduction in fringe cost
Less toolroom load due to less tooling required	25% less toolroom required	Ability to produce samples with production runs	50% of sample cost
		Availability of samples	A useful sales tool

Lower fixture cost—less needed	75% of durable fixture cost	Opportunity for foreman to concentrate on use of people rather than machines	Improved total operation
Less tool engineering time	30% of tool-process engineering cost	Reduction of direct labor	Actual savings based on pieces per week—not cycle time
Advantage of family-of-parts concept	20% of tool-process engineering cost		
Savings from less tool engineering—tool engineering records—tool drawings—process sheets, etc. (printing costs)	40% of printing cost	Flexibility of scheduling	Improved customer service
Machine maintenance savings due to improved and simpler designs	25% of machine repair—labor	Savings in scheduling	Improved flexibility
Less machine repair parts required	25% of machine repair—material	Ability to handle engineering changes	Simple program change
Less inspection due to improved machine—process repeatability	30% of inspection cost	Ability to handle variable raw material	Less raw material rejections
NC inspection more accurate than manual methods	Actual inspection time can be reduced as much as 80%	Ability to produce more complex parts	Machine capability simplifies tooling
Reduced setup time	80% of setup cost	Product engineering has more design flexibility	Can take advantage of NC capability
Reduced setup scrap	30% of scrap cost	Ability to handle future designs without extensive tooling	Program changes only will handle many new designs
Reduced scrap due to tool change or adjustment	20% of scrap cost	Reduces costs and improves estimating accuracy	Estimates can be dry run of tapes
More running time—80 to 85% versus 40 to 60%	10% of total burden	Skills built into tape programs retained through personnel changes	Tool and process engineers improved by 15%

SOURCE: K. M. Gettelman and M. D. Albert, eds., *Modern Machine Shop: 1982 NC/CAM Guidebook* (Cincinnati: Modern Machine Shop, 1982), pp. 434-35.

level and "diffusion" at a macroeconomic level resist easy conceptualization and measurement.[15] These processes are influenced by structural features of individual firms, the industry of which they are a part, internal and contextual aspects of both, and even characteristics of key actors who typically play roles of major importance in new and small businesses.[16]

In conformity with the popular model of industrial adjustment in which clusters of new industries displace older ones, great significance has come to be assigned to the role of innovation, entrepreneurship, and the small business sector. In one sense the recent lionization of small business has been a result of renewed infatuation with business formation, economic growth, and the personalities of highly visible risk takers as culture heroes. In another sense, however, the rekindled fascination with small business is rooted in a development that may be far more fundamental. Evidence is accumulating that goods-producing sectors in the United States are evolving in the direction of ever-smaller-scale production organizations. At a time when an estimated 75 percent of U.S. manufacturing is batch as opposed to mass or customized production, the role of small businesses in industrial adjustment acquires increased importance.[17] In industry after industry the trend toward smaller plants more articulated to the new scale features of modern manufacturing has been noted.[18] The relatively healthy small business sectors of low-growth and otherwise declining industries have been noted by Armington and Odle.[19] But the superiority of the small business sector in many of these respects has not yet been established unambiguously.[20] Legler and Hoy likewise note that the rate of increase in labor productivity has recently been slower for small firms than for larger firms.[21] It is at this point that the relation between the size and scale features of manufacturing and the diffusion and implementation of automated production technologies invites closer scrutiny.

Much current speculation suggests that the evolution and spread of advanced technologies in and through a sector are likely to proceed first and most easily in the newest—and thereby often the smallest—plants. The machine tool industry, for example, is dominated by very small companies; two-thirds have fewer than 20 employees, and less than 1 percent have more than 1,000. Yet small plants—defined as those with fewer than 1,000 workers—have been found to be more likely than large plants to acquire new metalworking equipment.[22] Furthermore, a recent inventory of metalworking equipment found that smaller plants employing between 20 and 99 employees have 35 percent of all the machine tools in the United States.[23] Moreover,

18

smaller plants generally had newer equipment than larger plants, despite the slow pace at which they have acquired automated machine control systems. In the future, therefore, increasingly automated production technologies may hold promise for making small companies more competitive with larger ones as well as compensating them for their difficulty in attracting and retaining skilled machinists, tool and die makers, and other machine operators.[24]

The model of industrial adjustment that guides this study and locates change within industries rather than between them similarly reserves for new and small business formations a significant role. An entire industry and even an entire sector can be slowly restructured to the extent that investment, output, employment, and productivity growth shift from older and larger to newer and smaller firms within it. As a result, many economic development initiatives in this decade have been formulated and implemented with such new small businesses in mind. Deregulation policies and tax subsidies for business growth, including even the reconceptualization of urban policy as tied closely to federal- and state-designated urban and rural enterprise zones, are cases in point. The aim has often been to capitalize on the fact that small businesses are more likely to remain dependent on and sensitive to their local environments and are thereby less likely to relocate.[25] Moreover, small businesses are seen as more willing to hire new labor force entrants and accommodate preferences for part-time work. As we shall see, however, in the search for simple categories by which to frame research and guide public policy responses, firm size may have been seized upon prematurely. Plant size—typically measured by size of work force—may be a poor proxy for important qualitative features that distinguish large plants from small ones or small plants from one another.

Innovation Investment by Plant Size. Speculation about the importance of small business to industrial change flows from our fascination in the past few years with the innovative capacity of new and small business in the American economy. The small business sector has been identified as the major wellspring of industrial innovations. According to a recent National Science Foundation report, small businesses were responsible for generating half of the most significant new industrial products and processes.[26] Moreover, they apparently do so efficiently as well. Even though small companies control less than 5 percent of the R&D dollars in the national economy, they are able to squeeze twenty-four times more major innovations from a dollar of R&D than large companies. Of the roughly 15,000 companies engaged in R&D, more than 90 percent have fewer than 1,000 employees; yet the less than 10 percent of firms with more than 1,000 employees did more than 90 percent of the R&D spending in 1979. When size is measured by assets or revenues, the superior innovativeness of smaller firms is once again apparent, even though larger firms might logically be thought to have the greater incentives and ability—revenues and cash flow—to channel resources to innovative activities.[27] Sharp has shown that smaller firms often spend proportionately more on innovative activities than larger ones.[28]

Employment Generation by Small Business. The role of small business in creating employment has also attracted considerable attention in recent years. In a study by Birch of 5.6 million businesses—which accounted for 80 percent of all private employment—relatively small establishments, with fewer than 250 employees, were reported to be responsible for creating 90 percent of the 6.8 million new jobs created between 1969 and 1976.[29] Two-thirds of these jobs were created by companies with fewer than 20 employees. And a majority of the jobs were created by establishments less than four years old.[30] The initial implication was that new small businesses created a disproportionately large share of new employment in relation to their share of the total economy. Armington and Odle have noted, however, that during 1978–1980 small business, defined as enterprises or firms (single-site or multilocational) with fewer than 100 employees, created a share of net new employment (37 percent) only slightly greater than their share of total U.S. employment (33 percent).[31] During 1980–1982 small businesses created 2.6 million new jobs, thus more than compensating for the loss by larger firms of 1.7 million existing jobs.[32] It has been estimated, however, that only 12–15 percent of small independent firms were responsible for this job creation.[33]

Small Business—Importance by Default? More recently the trendy image of small business as a significant employment generator has been sharply challenged. By implication the role of smaller enterprises as eager and willing adopters of sophisticated technologies is being called into question. At issue, it seems, is the "new economic religion in which small business plays savior for an ailing economy."[34] Bluestone and Harrison have been especially skeptical about the ability of the small business sector to lead this nation from an older to a more advanced industrial era. In their view the virtual dormancy of big business and its preoccupation with activities like acquisitions and mergers, widely regarded as nonproductive, have made the continuing yeasty growth of small and new business formations in the economy appear exceptional by default. Furthermore, while roughly half of all private sector jobs in the 1970s were generated by independent, single-site establishments, the growth rate in new facilities in components of multilocational corporations was much greater than it was for independents.

> While the "small business" sector is indeed large and visible—especially in an era of a domestic "capital strike" by the Fortune 500—it is still the corporate sector, and especially the smaller corporation, that is responsible for most of the jobs created by investment in new plants and stores—not independent entrepreneurs.[35]

The quality—as denoted by the fuller "social wage" tied to employment—as well as the quantity of jobs is likewise a test of value by which smaller enterprises often fail. Small businesses are often viewed as responsible for generating the low-wage, part-time, and minimum-benefit jobs that constitute the secondary labor market.

These factors, combined with the relatively high death rates of small business within the first five years, suggest that a preoccupation with the birth and diffusion of new small firms may be a weak reed upon which to base a policy-guided economic development, recovery, or later stabilization initiatives. Furthermore, on balance the evidence indicates that entrepreneurship is confounded with plant age and size at our own peril. While we should continue to value the bubbly growth in the new industrial sectors and the small businesses that generally lead in the development of those sectors, there is no justification for overlooking or underestimating the capacity of larger firms in older industrial sectors to renew and revitalize themselves.

Small Business Technology Adoption: A Special Case? Small businesses are different from other businesses because of their size, to be sure. Yet size may mask other, qualitative differences that are even more consequential in governing the dynamics of the production *of* innovations as well as production *with* innovations.[36] The substitution of "scale" for "size" has only restricted usefulness.[37]

The more an industry is dominated by small business, the greater its incentives may be to practice "offensive" innovation adoption, to demonstrate to larger firms that small business can adjust more swiftly and efficiently in competition. But the economies of scale available to large firms with their larger R&D budgets and research departments may help them retain advantages in most cases.[38] The special role of face-to-face communication in the transmission of ideas and information about new innovations and their uses also has a special relevance to the functioning of small business.[39]

While small businesses are typically more labor intensive, even when their productive output places them in the much-heralded high-technology sector, the combination of labor and various forms of production capital is continuously vulnerable to internal and external influences.[40] Nonetheless, modern production of goods and provision of services may both promote the adoption of upgraded production technologies and narrow the range of options available to a firm.[41] Thus the same feature that may influence the high death rate of new small businesses—productive efficiency—may also explain why innovation figures so prominently in the small businesses that survive. It becomes crucial, then, to identify and understand what barriers are faced by medium-sized and small businesses in adopting and implementing technological innovations to enhance their productivity, their competitiveness, and ultimately their chances for survival.

The choice of technology may be only very imperfectly guided by profit maximization since small businesses routinely overlook or abstain from adopting process technologies associated with demonstrated records of efficiency and the enhancement of productivity. "This raises the question of whether or not alternatives are in fact alternatives."[42] A crucial factor in NC/CNC-CAM adoption is the availability of technology with price and performance characteristics that reflect the scale and combination of tasks undertaken by small plants and shops, as well as the accessibility of the technology to the potential adopter. When smaller turnkey systems controlled by microcomputers became available at reduced prices, an entire new market was created among small firms facing narrow market niches. Whether an adoption-diffusion-implementation sequence is ultimately driven by demand or by supply, technology "fit" clearly needs to be more carefully explored and better understood.

Notes

1. International Trade Administration, *A Competitive Assessment of the U.S. Manufacturing Automation Equipment Industries* (Washington, D.C., 1984); and Business–Higher Education Forum, "The New Manufacturing: America's Race to Automate" (Washington, D.C., 1984), p. 2.

2. The geographical concentration of metalworking industries in older industrial regions may well have a variety of influences on the diffusion process, some of which may run counter to others. For example, Aldrich has suggested that the technology diffusion process in the machine tool industry was facilitated by its geographical concentration since the information crucial to new technology adoption could flow more easily within smaller-scale industrial networks. See H. E. Aldrich, *Organizations and Environments* (Engle-

wood Cliffs, N.J.: Prentice-Hall, 1979), p. 101; National Academy of Engineering, *The Competitive Status of the U.S. Machine Tool Industry* (Washington, D.C.: National Academy Press, 1983), p. 59; and International Trade Administration, *A Competitive Assessment*, p. 14.

3. For a similar forecast of rapidly expanding markets for automated manufacturing equipment, see Dun's *Business Month*, February 1984, p. C. Outlays for automated machine tools and controllers were projected to expand from $1.9 billion in 1967 and $3.9 billion in 1982 to $7.9 billion in 1987 and $18.0 billion in 1995, while outlays for all forms of factory automation were projected to exceed $37 billion by 1994.

4. See discussion of this trend in D. W. Austin and J. E. Beazley, "Struggling Industries in Nation's Heartland Speed Up Automation," *Wall Street Journal*, April 4, 1983.

5. The president of a firm producing CAD systems for printed circuit boards has reported his research results on the effect of price declines, which suggest that "only about 300 companies could afford a $350,000 system to design boards but some 6,000 could afford an $85,000 system." Cited in W. M. Bulkeley, "Computerized Design Systems Being Made for Smaller Firms," *Wall Street Journal*, November 12, 1982.

6. Given the signs of renewed interest in the purchase of machine tools, the sales of accessories for this shift from conventional to advanced manufacturing systems are expected to rise from $4 billion in 1982 to $30 billion worldwide by 1990. Data reported in Gene Bylinsky, "The Race to the Automated Factory," *Fortune*, February 21, 1983.

7. See Bulkeley, "Computerized Design Systems."

8. For a technical discussion of the origins and development of the family of numerically controlled automation technologies, see K. E. Gettelman, M. D. Albert, and W. Nordquist, "Introduction: Fundamentals of NC/CAM," in Gettelman et al., eds., *Modern Machine Shop: 1985 NC/CAM Guidebook* (Cincinnati: Modern Machine Shop, 1985), pp. 24–256. For a less technical but richly detailed discussion, see D. F. Noble, *The Forces of Production: A Social History of Industrial Automation* (New York: Knopf, 1984).

9. International Trade Administration, *A Competitive Assessment*, p. 5.

10. See R. J. Samuelson, "Business As Usual," *National Journal*, October 23, 1982, p. 1810.

11. For discussions of implementation and diffusion, see L. G. Tornatzky, J. D. Eveland, M. G. Boylan, W. A. Hetzner, E. C. Johnson, D. Roitman, and J. Schneider, *The Process of Technological Innovation: Reviewing the Literature* (Washington, D.C.: National Science Foundation, 1983), chaps. 7, 8.

12. Attempts to bring greater precision and conceptual clarity to the diffusion of managerial and technological innovations have been relatively numerous since the more generic diffusion literature originally attracted attention (E. M. Rogers with F. F. Shoemaker, *Communication of Innovations: A Cross-Cultural Approach* [New York: Free Press, 1971]). Aspects of diffusion that have been identified for separate and distinct examination include routinization (R. K. Yin, "Science and Technology in State and Local Governments: The Federal Role," in National Science Foundation, *The Five Year Outlook: Problems, Oppor-*

tunities, and Constraints in Science and Technology [Washington, D.C., 1980], vol. 2, pp. 649–61); institutionalization (J. D. Eveland, E. M. Rogers, and C. M. Klepper, *The Innovation Process in Public Organizations: Some Elements of a Preliminary Model*, Report to the National Science Foundation, Grant no. RDA 75-17952 [Ann Arbor: University of Michigan, 1977]); incorporation (W. H. Lambright, *Technology Transfer to Cities* [Boulder, Colo.: Westview Press, 1980]); stabilization (D. C. Pelz and D. Munson, "The Innovating Process: A Conceptual Framework," Working Paper, Center for Research on Utilization of Scientific Knowledge, University of Michigan, 1980); and continuation (G. Zaltman, R. Duncan, and J. Holbeck, *Innovations and Organizations* [New York: John Wiley and Sons, 1973]).

13. Figures reported in "The 13th American Machinist Inventory of Metalworking Equipment, 1983," *American Machinist* (November 1983), p. 1.

14. See International Trade Administration, *A Competitive Assessment*, table 10, p. 21.

15. J. D. Eveland, "Issues in Using the Concept of 'Adoption' of Innovations," *Journal of Technology Transfer*, vol. 4, no. 1 (1979), pp. 1–4; and Tornatzky et al., *Process of Technological Innovation*.

16. Pelz and Munson, "The Innovating Process"; T. J. Allen, *Managing the Flow of Technology* (Cambridge, Mass.: MIT Press, 1977); E. von Hipple, *The Role of the Initial User in the Industrial Good Innovation Process*, Final report to National Science Foundation (Cambridge, Mass.: Sloan School of Management, MIT, 1978); Eveland et al., *The Innovation Process*; Rogers and Shoemaker, *Communication of Innovations*; and R. T. Keller and W. E. Holland, *Technical Information Flows and Innovation Processes*, Final report to National Science Foundation (Houston, 1978).

17. Office of Technology Assessment, *Computerized Manufacturing Automation: Employment, Education, and the Workplace* (Washington, D.C., 1984), p. 37; and International Trade Administration, *A Competitive Assessment*, p. 1.

18. "Small Is Beautiful Now in Manufacturing," *Business Week*, October 22, 1984; and "Small Business Stumbles into the Computer Age," *Business Week*, October 8, 1984.

19. C. Armington and M. Odle, "Sources of Job Growth: A New Look at the Small Business Role," *Economic Development Commentary* (Council of Urban Economic Development) (Fall 1982), p. 5. See also "New Evidence on Small Business Role," *Socioeconomic Newsletter* (Institute for Socioeconomic Studies), vol. 8, p. 5 (August–September 1983).

20. K. H. Vesper, *Entrepreneurship and National Policy*, Heller Institute for Small Business Policy Paper, 1983, p. 31. Vesper has noted that "most small businesses are less productive than larger businesses when something other than a physical measure of output per unit of input is used."

21. J. Legler and F. Hoy, *Building a Comprehensive Data Base on the Role of Small Business* (Chicago: Heller Small Business Institute, 1982).

22. International Trade Administration, *A Competitive Assessment*, p. 26; and National Academy of Engineering, *Competitive Status*, p. 26.

23. "13th American Machinist Inventory."

24. Office of Technology Assessment, *Computerized Manufacturing Automation*, p. 136.

25. Armington and Odle, "Sources of Job Growth."

26. See also B. Bedell, "Aiding Small Business: Give the Money to the Real Innovators," *New York Times*, November 29, 1981. The observation that small firms have higher R&D productivity than large firms was originally noted by E. Mansfield, *Industrial Research and Technological Innovation: An Econometric Analysis* (New York: W. W. Norton, 1968). For an extensive review of the literature on the topic, see Tornatzky et al., *Process of Technological Innovation*, pp. 176ff.

27. D. C. Mueller, "A Life-Cycle Theory of the Firm," *Journal of Industrial Economics*, vol. 20 (July 1972), pp. 199–219.

28. M. Sharp, *The State, the Enterprise, the Individual* (New York: John Wiley and Sons, 1973). For a discussion of the ambiguity of the relation between firm size and innovation, see Aldrich, *Organizations and Environments*, pp. 103ff.

29. D. L. Birch, *The Job Generation Process* (Cambridge, Mass.: MIT Program on Neighborhood and Regional Change, 1979).

30. See N. R. Peirce and C. Steinbach, "Reindustrialization on a Small Scale—but Will the Small Businesses Survive?" *National Journal*, January 17, 1981, p. 105.

31. Armington and Odle, "Sources of Job Growth," p. 50.

32. Figures cited in *Business Week*, May 14, 1984.

33. Small Business Administration, *The State of Small Business: A Report to the President* (Washington, D.C., 1984), p. xv. Figure also cited in *Wall Street Journal*, February 6, 1985.

34. See R. J. Samuelson, "A False Religion," *National Journal*, November 20, 1982, p. 1992.

35. B. Bluestone and B. Harrison, *The Deindustrialization of America: Plant Closings, Community Abandonment, and the Dismantling of Basic Industry* (New York: Basic Books, 1982), p. 229.

36. Tornatzky et al., *Process of Technological Innovation*; T. D. Duchesneau, S. F. Cohn, and J. E. Dutton, *A Study of Innovation in Manufacturing: Determinants, Processes, and Methodological Issues*, Report to the National Science Foundation (Orono: University of Maine, 1979), vol. 1; J. R. Kimberly, "Organizational Size and the Structuralist Perspective," *Administrative Science Quarterly*, vol. 21 (1976), pp. 571–95; M. I. Kamien and N. L. Schwartz, "Market Structure and Innovation: A Survey," *Journal of Economic Literature*, vol. 13 (1975), pp. 1–37; and Kamien and Schwartz, *Market Structure and Innovation* (Cambridge: Cambridge University Press, 1982).

37. F. Hull and J. Hage, "A Systems Approach to Innovation and Productivity" (College Park: University of Maryland, Center for the Study of Innovation, 1981).

38. On oil refining and steelmaking, see B. Gold, W. S. Pierce, and G. Rosegger, "Diffusion of Major Technological Innovations in U.S. Iron and Steel Manufacturing," *Journal of Industrial Economics*, vol. 18 (1970), pp. 218–41;

and M. G. Boylan, "The Sources of Technological Innovation," in B. Gold, ed., *Research, Technological Change, and Economic Analysis* (Lexington, Mass.: Lexington Books, 1977). On chemical processes, see Kamien and Schwartz, "Market Structure and Innovation."

39. J. S. Coleman, E. Katz, and H. Menzel, *Medical Innovation: A Diffusion Study* (New York: Bobbs-Merrill, 1966).

40. W. A. Pasmore et al., *Sociotechnical Approaches to Organization Change in USAREUR* (Cleveland: Case Western Reserve University, 1980); P. R. Lawrence and J. W. Lorsch, *Organization and Environment* (Cambridge, Mass.: Harvard University Press, 1967); R. B. Duncan, "Characteristics of Organizational Environments and Perceived Environmental Uncertainty," *Administrative Science Quarterly* (1972), pp. 313–27; and H. K. Downey, D. Hellriegel, and J. W. Slocum, Jr., "Environmental Uncertainty: The Construct and Its Application," *Administrative Science Quarterly*, vol. 20 (1975), pp. 613–29.

41. N. Rosenberg, *Perspectives on Technology* (Cambridge: Cambridge University Press, 1976).

42. Tornatzky et al., *Process of Technological Innovation.*

3
The U.S. Metalworking Sector: An Overview

The larger aim of this study is to offer a deeper understanding of the trajectory of technology upgrading and how it can help metalworking industries adjust to new domestic and global economic circumstances.[1] Although the principal focus is on patterns of adoption of automated machine tool control systems by individual plants, this chapter describes the aggregate patterns of industrial change that have taken place in recent decades and against the background of which more specific adaptations, such as technology upgrading, must be viewed.

To study change in the U.S. metalworking sector is tantamount to studying this nation's adaptation to the full sweep of the industrial era itself. The dramatic evolution of the constituent industries in this sector over the past two centuries reflects not only the gradual domination of the U.S. economy by manufacturing but also the increased mechanization of production and its dependence on metals and their manipulation as a medium both for products and for the tools that basic manufacturing processes have come to require.

Today the metalworking sector is multifaceted and serves as an umbrella for a wide variety of disparate industries. Metalworking dominates the durable goods sector of manufacturing; it includes the primary and fabricated metal industries, the machinery and electronic equipment industries, and the transportation and instruments industries, among others. A common denominator for all these industries is that the medium for manufacture is metal, but long years of product and process diversification have created a highly variegated industry group. Moreover, metalworking industries dominate this nation's high-technology sector.

Yet this seeming diversity can conceal several enduring characteristics that suggest a sector at large that in some ways is very slow to change. Certain structural features have operated to constrain the modernization of the metalworking sector in recent decades. Reflecting its extended industrial heritage, many plants remain dominated by relatively static internal arrangements and outmoded physical capital. New ways of doing things embodied in new manufacturing technologies and more modern managerial arrangements have only marginally penetrated the operations of many plants in the sector. Furthermore, since the industry includes thousands of small plants and shops, either existing technologies have not been available on an appropriate scale to handle their workload tasks, or their uncertain productivity gains as against known costs have not been sufficiently attractive. In short, the rewiring of plants and shops and their workloads to accommodate automated production technologies throughout the sector has proceeded slowly at best.

Inevitably, the inertia in these background conditions evident at the plant level is often cited to explain why the man-machine interface on the shop floor has been slow to change in the direction of greater mechanization.[2] The longstanding power of skilled workers such as machinists in traditional manufacturing settings undergirds the relatively enduring structural features of the sector. As a result, in an era that assigns increasing importance to infusions of new technology into products and plant operations, much of the activity in the metalworking sector remains decidedly low tech.

Finally, the geographical concentration of these industries in older industrial regions—especially the East North Central and Middle Atlantic—is not easily overcome since the industry is tied to a dense network of suppliers, subcontractors, and industrial customers whose markets are also often highly localized. This pattern of concentration tends to amplify the consequences of intraregional competition and the job and market losses associated with industrial restructuring, business cycles, shifting trade patterns, and factory automation. Moreover, many metalworking industries are sensitive to the value-to-volume/weight ratios of their output, which are quickly translated into high transportation costs for plants and shops that work with metal. In a sense, then, the locational immobility of the industry is simply another reflection of the organizational inertia influencing the transformation of the production arrangements inside plants and shops.

The U.S. Metalworking Sector: A Setting for Change

In a strict sense the metalworking sector is neither a distinct nor a coherent industrial group. It is a conglomeration of plant and shop work settings in which are located machine shops whose principal activity involves performing mechanized metalworking operations.

Business establishments that bend, cut, or otherwise manipulate metal are the consumers of both conventional and automated machine tools manufactured by the relatively tiny, though strategically vital, machine tool industry.[3] The industry definition used in this study views the transformation of metal materials into finished or semifinished products as an economic activity first and foremost. It directs attention to the changing combinations of capital, labor, and other inputs that have defined metalworking over time and to the differing pace and direction of the substitutions among them at any one time.

While metalworking activities are also performed by manufacturing firms whose principal products are nonmetal (such as textiles or chemicals) as well as in nonmanufacturing settings, this share of total metalworking activity is trivial. The core of the U.S. metalworking sector includes more than 200 manufacturing industries at the four-digit SIC level that produce everything from automobiles and computers to beer cans and even the common pin immortalized by Adam Smith.[4]

The High-Technology Metalworking Sector

Commonly overlooked is the fact that many metalworking industries are among the nation's high-technology elite. Although the composition and size of the U.S. high-technology sector vary with the criteria used to define it, the sector is clearly dominated by manufacturing industries. If we classify industries as high technology when their share of technology-oriented workers is at least 1.5 times the average for all industries, thirty-eight of forty-eight high-technology industries, accounting for 68.4 percent of high-technology employment, were in manufacturing in 1982 (table 2).[5] Moreover, twenty-five metalworking industries accounted for 49.9 percent of the nation's high-technology employment.

The size and composition of the high-technology sector have been changing. Long-term structural shifts as well as shorter-term cyclical fluctuations are clearly evident in the data in table 2. High-technology employment expanded 26.0 percent from 1972 to 1980, only to contract 1.7 percent from 1980 to 1982. While high-technology services employment expanded 46.2 percent from 1972 to 1980, the growth in metalworking and nonmetalworking manufacturing was only 19.8 percent and 17.3 percent, respectively. Moreover, during the cyclical downturn in the early 1980s, metalworking industries sustained the bulk of the decline. From 1980 to 1982 employment declined in nineteen of the twenty-five metalworking industries, for a

29

TABLE 2
High-Technology Employment Trends, 1972–1982

Industry	SIC	Employment (thousands)			Percent Change	
		1972	1980	1982	1972–80	1980–82
Metalworking						
Ordnance and accessories	348	81.9	63.4	71.4	-22.6	12.6
Engines and turbines	351	114.6	135.2	114.8	18.0	-15.1
Farm and garden machinery	352	135.0	169.1	130.8	25.3	-22.6
Construction, mining, and material handling machinery	353	293.7	389.3	340.9	32.6	-12.4
Metalworking machinery	354	286.0	373.1	320.3	30.5	-14.2
Special industrial machinery, except metalworking	365	176.9	207.3	179.4	17.2	-13.5
General industrial machinery	356	267.5	323.7	283.2	21.0	-12.5
Office, computing, and accounting machines	357	259.6	432.2	489.7	66.5	13.3
Refrigeration and service industry machinery	358	164.4	174.2	161.3	6.0	-7.4
Electric transmission and distribution equipment	361	128.4	122.5	110.1	-4.6	-10.1
Electrical industrial apparatus	362	209.3	239.9	211.8	14.6	-11.7
Household appliances	363	186.9	163.2	142.0	-12.7	-13.0
Electric lighting and wiring equipment	364	204.4	209.2	186.9	2.4	-10.7
Radio and television receiving equipment	365	139.5	108.8	94.6	-22.0	-13.1
Communication equipment	366	458.4	541.4	555.7	18.1	2.6
Electronic components and accessories	367	354.8	553.6	568.7	56.0	2.7
Miscellaneous electrical machinery	369	131.7	152.1	141.3	15.5	-7.1
Motor vehicles and equipment	371	874.8	788.8	690.0	-9.8	-12.5

Aircraft and parts	372	494.9	652.3	611.8	31.8	-6.2
Guided missiles and space vehicles	376	92.5	111.3	127.3	20.3	14.4
Engineering, laboratory, scientific, and research instruments	381	64.5	76.8	75.7	19.1	-1.4
Measuring and controlling instruments	382	159.6	245.3	244.3	53.7	-0.4
Optical instruments and lenses	383	17.6	33.0	32.5	87.5	-1.5
Surgical, medical, and dental instruments	384	90.5	195.5	160.4	116.0	-18.0
Photographic equipment and supplies	386	117.1	134.6	138.3	15.0	2.7
Metalworking total		5,504.5	6,595.8	6,183.2	19.8	-6.3
Nonmetalworking						
Crude petroleum and natural gas	131	139.3	219.6	281.7	57.7	28.3
Heavy construction, except highway and street	162	495.1	658.5	633.9	33.0	-3.7
Industrial inorganic chemicals	281	141.2	161.1	153.5	14.1	-4.7
Plastic materials and synthetics	282	228.7	204.8	182.7	-10.5	-10.8
Drugs	283	159.2	196.1	199.8	23.2	1.9
Soaps, cleaners, and toilet preparations	284	122.4	140.9	145.3	15.1	3.1
Paints and allied products	285	68.6	65.1	59.7	-5.1	-8.3
Industrial organic chemicals	286	142.8	174.1	174.3	21.9	0.1
Agricultural chemicals	287	56.4	72.0	67.1	27.7	-6.8
Miscellaneous chemical products	289	90.0	93.3	91.5	3.7	-1.9
Petroleum refining	291	151.4	154.8	169.0	2.3	9.2
Tires and inner tubes	301	122.1	114.8	101.9	-6.0	-11.2
Cement, hydraulic	324	31.9	30.9	28.5	-3.1	-7.8
Nonmetalworking total		1,949.1	2,286.0	2,288.9	17.3	0.1

(Table continues)

31

TABLE 2 (continued)

Industry	SIC	Employment (thousands)			Percent Change	
		1972	1980	1982	1972–80	1980–82
Service						
Radio and television broadcasting	483	142.7	199.6	216.4	39.9	8.4
Communication services, n.e.c.	489	29.7	66.1	91.0	122.6	37.7
Electric services	491	312.0	391.0	415.1	25.3	6.1
Combination electric, gas, and other utility services	493	183.4	196.7	198.4	7.3	0.9
Wholesale trade, electrical goods	506	331.2	421.4	434.9	27.2	3.2
Wholesale trade, machinery, equipment, and supplies	508	868.6	1,307.7	1,344.9	50.6	2.8
Computer and data processing services	737	106.7	304.3	357.5	185.2	17.5
Research and development laboratories	7391	110.7	163.1	162.7	47.3	−0.2

	SIC					
Engineering, architectural, and surveying services	891	339.3	544.9	568.7	60.6	4.4
Noncommercial educational, scientific, and research organizations	892	111.8	113.5	117.8	1.5	3.8
Services total		2,536.1	3,708.3	3,907.4	46.2	5.4
High-technology total		9,989.7	12,590.1	12,379.5	26.0	-1.7
Metalworking sector share		55.1	52.4	49.9		
Nonmetalworking sector share		19.5	18.2	18.5		
Services share		25.4	29.5	31.6		

SIC = standard industrial classification; n.e.c. = not elsewhere classified.

NOTE: This high-technology sector includes industries whose share of technology-oriented workers (engineers, life and physical scientists, mathematical specialists, engineering and science technicians, and computer specialists) is at least 1.5 times the average for all industries.

SOURCE: Adapted from Richard W. Riche, Daniel E. Hecker, and John U. Burgan, "High Technology Today and Tomorrow: A Small Piece of the Employment Pie," *Monthly Labor Review* (November 1983), table 1, p. 52.

net job loss of 6.3 percent. In contrast, although the majority of the nonmetalworking industries lost employment, there was a slight net gain (0.1 percent); the services registered a 5.4 percent increase over the same period, with only one industry losing employment.

In summary, with high-technology industries defined by their dependence on clusters of special labor skills, the dominance of employment in the high-technology sector by metalworking industries is large, though declining. The metalworking manufacturing share of total high-technology employment declined steadily from 1972 to 1982 from 55.1 percent to 49.9 percent, while the nonmetalworking manufacturing share declined from 19.5 percent to 18.5 percent. The shift to high-technology services is apparent in their rise from a 25.4 percent to a 31.6 percent share over the same decade. As I have already noted, however, much of the productive activity in the metalworking industries is characterized by obsolescence and a resistance to change. To explore these seemingly contradictory circumstances more fully, let us examine in greater•detail some of the changes experienced by metalworking industries in recent years.

Establishment and Employment Change: 1980–1982

The employment contraction among metalworking industries in 1980–1982 was substantial. The composition and patterns of the employment and establishment changes are the focus of this section. The data reported are drawn from analyses of the U.S. Establishment Longitudinal Microdata (USELM) and the U.S. Enterprise and Establishment Microdata (USEEM) files developed by the Brookings Institution and maintained by the U.S. Small Business Administration.[6]

Patterns of Establishment Change. Plants and shops, whether tied to single-site or to multiestablishment enterprises, are the settings for industrial change of all kinds, including employment change. Multiestablishment enterprises constituted approximately 12 percent of the USEEM files in 1982. This section offers an overview of the changes experienced by plants and shops in the metalworking sector during 1980–1982. The total number of metalworking establishments increased 1.41 percent during that period (table 3). This expansion exceeded that for all manufacturing (0.12 percent) but lagged behind that for all industries (2.60 percent). The largest net addition of plants came in the nonelectrical machinery industry, although its rate of plant expansion was matched by that of the instruments industry. The industry groups tied to steel, automobiles, and other forms of metal fabrication lost plants during the period.

34

Plant formations were 10 percent higher than plant failures, adding slightly more than 3,000 metalworking establishments during 1980–1982. This change is the result of a larger process in which the loss of more than 30,000 establishments through business failure was barely compensated for by the gain of nearly 34,000 through new business formation. Moreover, the rate of net plant expansion compares favorably with that for all manufacturing after the failure and formation of nearly 70,000 manufacturing establishments. The ratios of plant formation to failure were highest in the nonelectrical machinery industry (1.31) and the instruments and related products industry (1.22); they were lowest for the industry dominated by steel (0.87).

Patterns of Employment Change. From all indications, the metalworking industry experienced tumultuous and widespread employment contraction as well as turnover among establishments during the early 1980s. A considerable portion of this employment change was tied to turnover at metalworking plants and shops. Data from the USEEM file show that metalworking employment stood at 12.6 million in 1982, down 3.8 percent from 1980 (table 4). Even though metalworking industries sustained widespread and substantial losses during that time, however, the rate of job loss in total manufacturing was even more severe—5.2 percent. As a result, one effect of the cyclical downturn during this period was that metalworking's share of total manufacturing employment actually increased slightly, from 54.0 percent to 54.9 percent, while its share of total employment declined from 15.4 percent to 14.6 percent. The bulk of metalworking employment is in electrical and nonelectrical machinery, transportation equipment, and fabricated metal products. Employment loss was widespread, with the electric and electronic equipment and instrument industries alone fending off employment contraction. Not surprisingly, the primary metal industries—especially the steel industry—sustained the most severe job losses during this period.

Net employment shifts may mask considerable underlying changes tied to the turnover of new business formations and failures among existing businesses, as well as hiring and layoffs in existing plants. Table 4 indicates that considerable employment activity took place beneath the surface of net employment changes. The ratio of metalworking jobs gained to jobs lost for whatever reason was 0.82. The efficiency with which jobs gained compensated for jobs lost in metalworking industries is indicated by the lower ratio of jobs gained to jobs lost in total manufacturing employment (0.75).

Nearly 3 million metalworking jobs were lost during 1980–1982. More jobs were lost because metalworking plants went out of busi-

TABLE 3
COMPONENTS OF ESTABLISHMENT CHANGE, 1980–1982

Industry	SIC	Total Establishment Change, 1980–1982				Establishment Change Due to Plant Turnover			
		1980	1982	Net change	Percent change	Plant formation	Plant failure	Net change	Ratio of plant formation to failure
Furniture and fixtures	25	15,515	15,515	—	0.00	2,447	2,447	—	1.00
Primary metal industries	33	10,889	10,669	−220	−2.02	1,410	1,630	−220	0.87
Fabricated metal industries	34	47,065	46,862	−203	−0.43	5,675	5,878	−203	0.97
Machinery, except electrical	35	74,430	77,017	2,587	3.48	10,829	8,243	2,586	1.31

Electric and electronic equipment	36	24,860	25,399	539	2.17	4,676	4,138	538	1.13
Transportation equipment	37	13,521	13,306	−215	−1.59	2,200	2,415	−215	0.91
Instruments and related products	38	13,568	14,017	449	3.31	2,520	2,072	448	1.22
Miscellaneous manufacturing	39	24,416	24,537	221	0.91	3,788	3,567	221	1.06
Total U.S. metalworking establishments		224,264	227,422	3,158	1.41	33,545	30,390	3,155	1.10
Total U.S. manufacturing establishments		463,853	464,388	535	0.12	67,212	66,677	535	1.01
Total U.S. establishments		5,042,114	5,172,973	130,859	2.60	810,328	679,469	130,859	1.19

SOURCE: Adapted from U.S. Small Business Administration, Office of Economic Research, USEEM 1980–1982 longitudinal weighted data.

TABLE 4

EMPLOYMENT GAINED AND LOST BY FORMATION, FAILURE,
RELOCATION, EXPANSION, AND CONTRACTION OF
PRIVATE METALWORKING ESTABLISHMENTS IN THE UNITED STATES,
1980–1982

Industry	Total Employment Change, 1980–1982			
	1980	1982	Net change	Percent change
Furniture and fixtures	606,616	571,793	−34,823	−5.74
Primary metal industries	1,431,792	1,287,285	−144,507	−10.09
Fabricated metal industries	1,999,320	1,854,595	−144,725	−7.24
Machinery, except electrical	3,172,516	3,132,004	−40,512	−1.28
Electric and electronic equipment	2,438,384	2,465,569	27,185	1.11
Transportation equipment	2,026,133	1,917,416	−108,717	−5.37
Instruments and related products	787,779	789,080	1,301	0.17
Miscellaneous manufacturing	624,122	576,537	−47,585	−7.62
Total U.S. metalworking employment	13,086,662	12,594,279	−492,383	−3.76
Total U.S. manufacturing employment	24,216,816	22,952,054	−1,264,762	−5.22
Total U.S. employment	85,117,539	86,105,790	988,251	1.16

TABLE 4 (continued)

| | Employment Change Due to Plant Turnover | | | |
Industry	Plant formation	Plant failure	Net change	Ratio of formation to failure
Furniture and fixtures	48,307	80,951	−32,644	0.60
Primary metal industries	93,329	200,646	−107,317	0.47
Fabricated metal industries	138,744	239,154	−100,410	0.58
Machinery, except electrical	273,683	310,215	−36,532	0.88
Electric and electronic equipment	204,254	273,104	−68,850	0.75
Transportation equipment	113,646	178,436	−64,790	0.64
Instruments and related products	72,186	86,549	−14,363	0.83
Miscellaneous manufacturing	45,020	86,244	−41,224	0.52
Total U.S. metalworking employment	989,169	1,455,299	−466,130	0.68
Total U.S. manufacturing employment	1,756,806	2,961,083	−1,204,277	0.59
Total U.S. employment	9,333,767	11,137,691	−1,803,924	0.83

(Table continues)

39

TABLE 4 (continued)

| Industry | Employment Change within Existing Plants | | | | |
	Expansion	Contraction	Net change	Ratio of expansion to contraction	Ratio of Total Jobs Gained to Jobs lost
Furniture and fixtures	56,556	58,736	−2,180	0.96	0.75
Primary metal industries	84,502	121,692	−37,190	0.69	0.55
Fabricated metal industries	149,538	193,853	−44,315	0.77	0.67
Machinery, except electrical	330,357	334,337	−3,980	0.99	0.94
Electric and electronic equipment	313,400	217,365	96,035	1.44	1.06
Transportation equipment	170,943	214,871	−43,928	0.80	0.72
Instruments and related products	98,719	83,055	15,664	1.19	1.01
Miscellaneous manufacturing	55,376	61,737	−6,361	0.89	0.68
Total U.S. metalworking employment	1,259,391	1,285,646	−26,255	0.98	0.82
Total U.S. manufacturing employment	2,134,276	2,194,761	−60,485	0.97	0.75
Total U.S. employment	10,206,665	7,414,490	2,792,175	1.38	1.05

SOURCE: Adapted from U.S. Small Business Administration, Office of Economic Research, USEEM 1980–1982 longitudinal weighted data.

ness or relocated than were lost by layoffs in existing plants, but each kind of job loss claimed well over 1 million jobs. It is especially noteworthy that job loss through business failure and relocation did not overwhelm job creation through new business formations and relocations uniformly among all metalworking industries. More than 2.2 million new jobs were added, the bulk of them in existing plants. Nearly 1 million new metalworking jobs were gained from new plant establishment and relocation, but the simultaneous loss of nearly 1.5 million jobs through business failure and relocation resulted in a net loss of nearly half a million metalworking jobs through plant turnover. The ratio of jobs gained to jobs lost because of business formation, failure, and relocation was 0.68. By contrast, in existing plants and shops, new hires nearly equaled layoffs, the ratio of jobs gained to jobs lost being 0.98. Thus far less of the net employment change in metalworking during 1980–1982 was tied to shifting employment levels in existing plants than to plant turnover.

Small Metalworking Business Trends

The broad base of the metalworking industry is composed of thousands of small plants and shops, which are of particular importance in this study of industrial adjustment and technology upgrading. With small businesses defined here as those employing fewer than 250 employees, individual industries vary considerably in the extent to which their ranks are dominated by such businesses.[7] Nonetheless, the employment and sales shares accounted for by small metalworking businesses are roughly comparable. Recessions often affect different segments of an industry differently, however, and thus alter the role of small business. Table 5 reports data on trends in employment and sales shares experienced by metalworking firms and eventually registered as changes in the shares accounted for by the small business sector in metalworking. The data are for both single-site and multiestablishment enterprises.[8]

Small Business Employment Shares. Employment share trends for small metalworking businesses vary considerably by sector. In 1982 more than half of total employment in standard industrial classifications (SIC) 25, 34, and 39 was in small establishments. Among enterprises the role of small business varied from relatively insignificant in SIC 37 to substantial in SIC 39 and 25. These patterns, however, do not diminish the fact that in all industries large businesses account for the bulk of employment by enterprise. This is especially true for enterprises in the automobile, steel, and electrical equipment industries.

TABLE 5

SMALL BUSINESS SHARE OF THE U.S. METALWORKING SECTOR, 1980–1982
(percent)

Industry	SIC	By Enterprise			By Establishment		
		1980	1982	Percent change 1980–82	1980	1982	Percent change 1980–82
		Total employment shares					
Furniture and fixtures	25	45.4	39.3	−13.4	59.1	59.5	0.7
Primary metal industries	33	12.0	13.8	15.0	54.3	29.5	−45.7
Fabricated metal products	34	38.5	37.4	−2.9	61.2	57.8	−5.6
Nonelectrical machinery	35	23.9	25.0	4.6	42.9	39.0	−9.1
Electric and electronic equipment	36	12.6	13.0	3.2	28.3	25.7	−9.2
Transportation equipment	37	5.1	7.5	47.1	26.0	13.6	−47.7
Instruments and related products	38	16.6	16.1	−3.0	28.0	31.0	10.7
Miscellaneous manufacturing industries	39	45.2	46.7	3.3	70.3	65.0	−7.5

Total sales shares

Furniture and fixtures	25	44.6	45.1	1.1	58.2	68.2	17.2
Primary metal industries	33	10.4	12.2	17.3	55.0	40.7	-26.0
Fabricated metal products	34	37.4	38.4	2.7	60.2	61.0	1.3
Nonelectrical machinery	35	21.1	19.8	-6.2	38.8	39.7	2.3
Electric and electronic equipment	36	12.0	10.7	-10.8	25.2	23.1	-8.3
Transportation equipment	37	3.9	4.3	10.3	19.6	23.2	18.4
Instruments and related products	38	13.9	13.1	-5.8	23.7	25.7	8.4
Miscellaneous manufacturing industries	39	44.8	44.2	-1.3	70.8	71.1	0.4

SOURCE: Adapted from U.S. Small Business Administration, Office of Economic Research, USEEM 1930–1982 data.

Over time, however, the role of small business has increased in a majority of these industries. Table 5 shows widespread, if modest, increases in the employment shares of small firms over the same period when many middle-sized firms were forced to cut back their employment and consequently tumbled down into the ranks of small business.

At the level of individual establishments, the role of small business takes on more significance. By 1982, partly because of the effect of back-to-back recessions on the fortunes of networks of metalworking plants tied together either formally through complex corporate structures or functionally through parts supply and related subcontracting links, the small business share of metalworking employment had contracted in all but two industries. This pattern is especially evident for those businesses closely linked to the automobile and steel industries. In sectors dominated by steel (-45.7 percent) and automobile (-47.7 percent) production, the share of total employment located in small plants was nearly halved during 1980–1982 as larger plants showed a greater capacity to weather the cyclical downturn.

Small Business Sales Shares. With respect to total sales, the dominant role of large businesses in several industries is once again apparent. In half the industries, however, the share of total sales accounted for by small enterprises increased during 1980–1982. This was especially evident in the steel industry complex, in which large integrated steel companies revealed a rapidly declining capacity to compete with minimills oriented to smaller, yet rapidly growing, market niches. The rising fortunes of small firms in the automobile industry complex are also indicated by the data. In the electrical equipment industry, however, an increasing share of total sales was derived from large firms. At the level of establishments, the significance of small plants and shops increased in all but two industries. The substantial decline in sales shares in primary metals establishments merits special attention. It is apparent that the effect of the recession on this industry was complex. Not only did small enterprises come to capture a larger share of total sales by 1982, but relatively larger establishments did likewise. This pattern reveals the especially severe effects of the recession and long-term declines in demand on small plants and may be due partly to the frequency with which larger corporations in this complex closed branch plants during the recession.

Employment Change by Employment Size of Plant. The role of small business in the metalworking sector is variable, although it is commonly enhanced during cyclical downturns. To what extent did the

44

TABLE 6

Employment Changes and Shifts in Employment Share by Establishment Size, U.S. Metalworking Sector, 1980-1982

Industry	SIC	Establishment Size	Employment Change 1980–82 (%)	Share of Total Industry Employment (%)	
				1980	1982
Furniture and fixtures	25	Less than 20	15.67	9.25	11.36
		20–99	−1.39	17.89	18.72
		Less than 100	4.43	27.15	30.07
		100–499	−10.32	22.96	21.85
		Less than 500	−2.33	50.11	51.92
		500+	−9.16	49.89	48.08
		Total	−5.74		
Primary metal industries	33	Less than 20	20.14	1.92	2.56
		20–99	−1.41	6.40	7.01
		Less than 100	3.56	8.32	9.58
		100–499	−11.41	9.94	9.79
		Less than 500	−4.59	18.25	19.37
		500+	−11.32	81.75	80.63
		Total	−10.09		
Fabricated metal products	34	Less than 20	14.49	8.77	10.82
		20–99	−3.01	18.89	19.75

(Table continues)

TABLE 6 (continued)

Industry	SIC	Establishment Size	Employment Change 1980–82 (%)	Share of Total Industry Employment (%)	
				1980	1982
		Less than 100	2.54	27.66	30.58
		100–499	−6.59	17.50	17.62
		Less than 500	−1.00	45.16	48.20
		500+	−12.38	54.84	51.80
		Total	−7.24		
Machinery, except electrical	35	Less than 20	16.27	9.47	11.16
		20–99	−0.59	13.06	13.15
		Less than 100	6.50	22.53	24.31
		100–499	−6.02	11.18	10.65
		Less than 500	2.35	33.72	34.95
		500+	−3.12	66.28	65.05
		Total	−1.28		
Electrical and electronic equipment	36	Less than 20	34.04	3.09	4.09
		20–99	7.76	7.17	7.64
		Less than 100	15.67	10.26	11.73
		100–499	−2.79	10.59	10.18
		Less than 500	6.29	20.84	21.91
		500+	−0.25	79.16	78.09
		Total	1.11		

Transportation equipment	37	Less than 20	17.45	2.12	2.63
		20–99	-2.23	4.24	4.39
		Less than 100	4.31	6.37	7.02
		100–499	-5.35	5.86	5.86
		Less than 500	-0.32	12.22	12.88
		500+	-6.07	87.78	87.12
		Total	-5.37		
Instruments and related products	38	Less than 20	30.25	5.81	7.56
		20–99	3.71	9.64	9.98
		Less than 100	13.69	15.45	17.54
		100–499	1.21	12.15	12.28
		Less than 500	8.20	27.61	29.82
		500+	-2.90	72.39	70.18
		Total	0.17		
Miscellaneous	39	Less than 20	15.30	15.34	19.15
		20–99	-5.90	20.85	21.24
		Less than 100	3.09	36.19	40.39
		100–499	-13.12	19.80	18.62
		Less than 500	2.64	55.99	59.01
		500+	-13.96	44.01	40.99
		Total	-7.62		

(*Table continues*)

TABLE 6 (continued)

Industry	SIC	Establishment Size	Employment Change 1980–82 (%)	Share of Total Industry Employment (%)	
				1980	1982
U.S. metalworking total		Less than 20	18.34	6.25	7.70
		20–99	−0.67	11.15	11.51
		Less than 100	6.16	17.41	19.20
		100–499	−6.55	12.09	11.74
		Less than 500	0.95	29.50	30.95
		500+	−5.74	70.50	69.05
		Total	−3.76		

SOURCE: Adapted from U.S. Small Business Administration, Office of Economic Research, USEEM 1980–1982 data.

recession have differing effects on employment in plants of different employment sizes? A partial answer to this question is to be found in the shifting employment shares of plants of different sizes during 1980–1982. The role of firm size—as measured by employment at the establishment level—in insulating firms of certain sizes from employment changes and thereby channeling such changes among firms of different sizes is the focus here. The share of total employment in firms with fewer than 500 employees increased during 1980–1982 for the entire metalworking sector, while the employment share in larger firms declined (table 6). This pattern was replicated for firms with fewer than 100 employees in each of the two-digit metalworking subsectors, but this shift to small business was mainly confined to firms employing fewer than 100. For a majority of industries, the employment share of plants employing 100 to 500 declined.

In summary, it is apparent that large firms and large-scale production arrangements still dominate much of U.S. metalworking. Nonetheless, changes may be under way that could in effect decouple corporate organization from scale of production. Large and even multilocational firms may well be in the process of rescaling their production activities to accommodate the shift to more specialized batch production runs. As a result, both longer-term structural shifts through advanced industrial evolution and the shorter term cyclical effects of business recessions magnify the importance of smaller-scale and small-business-based manufacturing. These and related industrial adjustments, to the extent that they continue, have the capacity both to facilitate and to hinder the diffusion and implementation of advanced manufacturing technology. It is to the range of longer-term structural shifts, which can slowly reconfigure an entire industry as a target for the diffusion of new technology, that I now turn.

Notes

1. Lawrence has illustrated the greater importance of the restructuring of domestic demand than of rising imports in explaining the employment contraction in the domestic automobile and steel industries. See R. Z. Lawrence, *Can America Compete?* (Washington, D.C.: Brookings Institution, 1984), pp. 54ff.

2. The staying power of traditional forms of shop floor organization involving machinists and their machines is evident in plants throughout the metalworking sector. This enduring feature of the sector appears to challenge the conventional ecological view that the industrial adjustment process at the aggregate level can be expected to select out plants and shops whose internal organizational features, though increasingly outmoded, are inflexible in the face of changing economic and technological conditions. Instead, this sector

offers evidence of "structural inertia" and the "liability of newness" wherein certain forms of rigidity tied to characteristics of metalworking plants and their economic environment have survival value. See M. T. Hannan and J. Freeman, "Structural Inertia and Organizational Change," *American Sociological Review*, vol. 49 (April 1984), pp. 149–64; J. Freeman, G. R. Carroll, and M. T. Hannan, "The Liability of Newness: Age Dependence in Organizational Death Rates," *American Sociological Review*, vol. 48 (October 1983), pp. 692–710; and A. L. Stinchcombe, "Social Structure and Organizations," in J. G. March, ed., *Handbook of Organizations* (Chicago: Rand McNally, 1965), pp. 153–93. I am indebted to my colleague Larry Redlinger for helping me to reappreciate this perspective on organizational change more fully. See also W. Dostal et al., "Flexible Manufacturing Systems and Job Structures," Mitteilungen aus der Arbeitsmarkt und Berufsforschung, 1982; and W. Dostal and K. Kostner, "Changes in Employment with the Use of Numerically Controlled Machine Tools," Mitteilungen aus der Arbeitsmarkt und Berufsforschung, 1982, cited and reviewed in Office of Technology Assessment, *Computerized Manufacturing Automation: Employment, Education, and the Workplace* (Washington, D.C., 1984), chap. 4.

3. The machine tool industry combines metal-cutting (SIC 3541) and metal-forming (SIC 3542) machine tool manufactures.

4. Adam Smith, *The Wealth of Nations* (New York: Modern Library, 1937; first edition, 1776). It is estimated by the research staff at *Modern Machine Shop* that 96.8 percent of the businesses with machine shops were in manufacturing settings in 1984; 93.4 percent were located in industries whose outputs were fashioned from metal (SIC 25, 33–39) and only 3.4 percent in industries whose outputs were not metal (SIC 20–24, 26–32). The remaining 3.2 percent of machine shops were in government, educational, and durable goods wholesale settings and metal services centers (*Modern Machine Shop*, Cincinnati, Ohio).

5. The "high-technology" sector is constituted according to the criteria developed by the Bureau of Labor Statistics and employed in Richard W. Riche, Daniel E. Hecker, and John U. Burgan, "High Technology Today and Tomorrow: A Small Piece of the Employment Pie," *Monthly Labor Review* (November 1983), pp. 50–58.

6. Generally, establishment and employment figures available from the *Census of Manufactures* are substantially lower than those from the USELM or USEEM files. Members of the Small Business Administration research staff assert that the Dun and Bradstreet files from which both files have been constructed provide better coverage than those based on government economic surveys. Although the employment and establishment numbers in the USELM and USEEM files differ from those of the Census Bureau, the change data are generally comparable. Unless otherwise noted, the data reported in this section are based on the longitudinal weighted files, in which enterprise employment and the sum of establishment employment have been statistically reconciled through regression techniques. Since employment totals at the enterprise level are viewed as more reliable than the sum of establishment totals within an enterprise, the USELM file employs a reconciliation process to

adjust employment levels for establishments against enterprise totals. In the process, approximately 6 million foreign employed have been removed from the overall file. For a fuller explanation of this longitudinal weighted and adjusted file, see B. D. Phillips, "A Guide to Understanding U.S. and State Small Business Job Generation Data," U.S. Small Business Administration, Office of Advocacy, December 1984.

7. The most common employment criteria used to classify a firm as a small business are 100 and 500 employees. The analyses reported in the remaining chapters of this study define enterprises and establishments with 250 or fewer employees as small businesses.

8. Since employment and sales data by employment size of establishment were not available in the weighted USELM data file, unweighted data are reported in table 5.

4

Varieties of Industrial Adjustment

The primary intent of this chapter is to offer a strategy for exploring the larger process of industrial adjustment in so-called mature manufacturing industries. The broad base of the U.S. metalworking sector is viewed as the target of a new surge of adoptions of automated manufacturing technologies, a diffusion long hampered by the small size and scale of production of the firms involved and the poor fit between their production tasks and shop floor arrangements and the availability and affordability of advanced production technologies. But the adoption by plants and shops of advanced machine control technologies such as NC/CNC-CAM is only one of many ways in which an industry can be gradually repositioned in an ever-changing economic environment. The upgrading of production technologies as an industrial adjustment strategy must be seen against the backdrop of important longer-term and unorchestrated industry adjustments whose consequences may either hinder or facilitate their adoption.

This chapter presents evidence of significant industrial adjustments at three conceptual levels to help us better understand the patterns of technology upgrading reported in chapter 5. All three adjustments mirror the continuing development of the larger metalworking sector. The first is the shifting pattern of geographical concentration. As regions (and census divisions within them) experience differing capacities to incubate new firms or retain existing ones, the critical mass of an industry may gradually migrate from certain regions—and the business environments they provide—to others.[1] The mobility of capital, quite apart from the physical relocation of existing plants, is at the heart of this process.

The second adjustment is tied to the complexity of the corporate structure of metalworking industries. The links between a firm's headquarters and its satellite production and distribution centers have emerged as important features of industrial development in manufacturing in the past half-century.[2] My goal here is to trace this corporate structure in the metalworking firms in my sample. Finally, the third adjustment concerns features of the work setting inside the plant and how they may be influenced by whether or not production workers are unionized. The potential for an industry to exchange one labor environment for another as a result of different growth rates of union and nonunion plants in right-to-work and non-right-to-work states is explored. Before considering these examples of industrial adjustment in the U.S. metalworking sector, however, I first consider the general process of industrial adjustment in more detail.

Industrial Adjustment: Plant Adaptation and Turnover

The importance and inevitability of a full range of industrial adjustments involving labor, physical capital, energy, technology, and related production inputs appear to have been rediscovered and vigorously promoted in recent years. In the U.S. metalworking industry complex, however, adjustment has been a continuous process, occurring at several levels throughout this century. As in all processes of industrial change, industrial organization may accommodate a changing economic environment and continuing technological development, as well as initiate and sustain changes of both kinds in two major ways. The first consists of adaptations by existing plants and shops through expansion, contraction, and reorganizing or rationalizing key production activities in response to changing circumstances. The second is the changing size and corporate composition of the industry, which reflects the net effects of a process of plant replacement as a succession of business formations and failures continuously reconfigures the industry.

In the first process, industrial adjustment is viewed as a series of adaptations of workers and work settings to changing conditions in the larger economic environment. From this perspective industrial adjustment is a complex phenomenon that may proceed in disparate and subtle ways. Individual plants and shops seldom face uniform environmental circumstances or have comparable capacities to respond. Not unexpectedly, therefore, the adaptations plants make—or fail to make—vary considerably. As plants and shops respond in different ways, to different degrees, at varying speeds, and with differing effects, the cumulative result is inevitably reflected in the larger industry. Accelerated or deferred capital investment in new plant and equipment or the shift of production to satellite plants in search of lower-cost business settings are simply two examples of the innumerable adjustments that existing plants and shops may make.

Industrial adjustments made by existing firms take place in the context of a process of continuous plant turnover on a larger scale. As new cohorts of plants are established and older ones go out of business, an industry may be substantially restructured. In the process,

53

outmoded and uncompetitive production arrangements embodied in older plants are often replaced by newer and more competitive ones in newer plants. While we must be cautious about assuming that the age of a plant, the vintage and sophistication of its capital stock, and its competitiveness are always highly correlated, the process of cohort succession-replacement does offer an avenue for the continual upgrading of an industry. As a result, a number of features of an industry may fluctuate while others remain stable. As a consequence of differing rates of industrial development among regions, for example, the production plants of an industry may be geographically dispersed while administrative control remains concentrated in corporate headquarters.[3] Or small firms may be established in record numbers while larger firms continue to account for the largest shares of total output. These are but two of countless ways in which the organization of an industry within and among plants may gradually be shaped by the economic environment. I turn now to examples of shifts that illustrate the gradual and continuous adjustment of the U.S. metalworking industry complex.

The U.S. Metalworking Sector: Background Characteristics

Measuring technological effects, including all aspects of diffusion, on the broad base of the metalworking sector, composed of hundreds of industries and thousands of small plants and shops, is central to this study. A special data base was used for most of the analyses below. A data base composed of plants housing metalworking machine shops was created and has been maintained for over half a century by the research staff of *Modern Machine Shop*. In October 1982 approximately 38,600 plants and shops with fewer than 250 employees and 12,523 plants and shops with at least minimal NC/CNC machine tool control capabilities existed in the United States. For this study, a universe of 7,994 of these plants and shops with at least minimal automated machine tool control capabilities was defined. A mail survey was conducted to gather information from plants in this universe. Further information on the research design is presented in the appendix.

State and Regional Plant Location. One indication of the representativeness of the sample is that the geographical distribution of plants in the universe is closely replicated in the subset of plants on which the analyses below are based. Both distributions are presented in table 7.[4] The greatest concentrations of metalworking plants are in the East North Central (32.2 percent) and Middle Atlantic (19.7 percent) regions. These plants are concentrated in New York, Pennsylvania,

Ohio, Illinois, and Michigan in the East and Midwest and California in the Far West.

The plant locations reveal the way in which the contours of the original industrial heartland have been adjusted as the metalworking industry complex has developed and markets and production facilities have dispersed. Although these locational patterns have been developing for more than two centuries, the same basic geographical distribution was already well established over a century ago. To a great extent, this patterning reflects the resource and market dependencies of the manufacturing sector and key components of the metalworking complex, including especially the relatively small machine tool industry. In a similar fashion, the tendency for metalworking firms to be owned and managed by individuals or families has brought about an inertia, in which expansion is accommodated at an original location or at a satellite location nearby.[5] The concentrations of plants in the older Northeast and Midwest regions and the rise of new concentrations after World War II, particularly in California, illustrate a system of interdependencies among metalworking firms, their suppliers, and their customers in metal products industries that is as organizationally complex as it is locationally stable.

The state and regional concentrations that define the metalworking sector indicate backward and forward linkages that may be shaped as much by characteristics of the products as by those of the firms themselves. The relatively low value-to-volume/weight ratios that characterize many metal product markets reinforce the historical concentration of metalworking activities. Rival suppliers of metal components that try to compete from locations outside these invisible and shifting economic boundaries must absorb relatively high transport costs. As those costs are passed along, they can elevate prices to levels that render intermediate inputs and finished products uncompetitive. Despite the development of lighter-weight metals such as aluminum, titanium, and related nonferrous alloys, the volume/weight factor continues to preserve the relatively local relations among metalworking firms.

Plant "Birth" Cohorts and Age of Physical Plant. The date when production began at a plant is often used as an indicator, though an imperfect one, of the age of capital plant and the sophistication of the production equipment inside. Yet this can be quite misleading. A key feature of the diffusion of technology is the way in which an older physical plant may be upgraded by retrofitting it with advanced production and support technologies, including automated machine control systems like NC and CNC, robotics, computer-assisted design

TABLE 7

STATE AND REGIONAL DISTRIBUTION, PLANT UNIVERSE AND SAMPLE, 1982

	Universe		Sample	
	No.	Percent	No.	Percent
New England	790	9.9	108	9.2
Maine	25		4	
New Hampshire	60		11	
Vermont	14		2	
Massachusetts	390		54	
Rhode Island	37		7	
Connecticut	264		30	
Middle Atlantic	1,571	19.7	213	18.2
New York	596		80	
New Jersey	391		44	
Pennsylvania	584		89	
East North Central	2,574	32.2	388	33.1
Ohio	762		111	

	Universe		Sample	
	No.	Percent	No.	Percent
South Carolina	55		6	
Georgia	48		7	
Florida	150		24	
East South Central	193	2.4	25	2.1
Kentucky	63		9	
Tennessee	71		10	
Alabama	47		4	
Mississippi	12		2	
West South Central	481	6.0	70	6.0
Arkansas	29		6	
Louisiana	33		6	
Oklahoma	89		12	
Texas	330		46	

	Number	Percent	Number	Percent
Indiana	241		36	
Illinois	711		103	
Michigan	573		85	
Wisconsin	287		53	
West North Central	621	7.8	108	9.2
Minnesota	219		45	
Iowa	80		13	
Missouri	188		20	
North Dakota	4		2	
South Dakota	17		3	
Nebraska	30		6	
Kansas	83		19	
South Atlantic	546	6.8	71	6.1
Delaware	4		0	
Maryland	67		11	
Washington, D.C.	1		0	
Virginia	67		6	
West Virginia	24		2	
North Carolina	130		15	
Mountain	221	2.8	38	3.2
Montana	1		0	
Idaho	3		0	
Wyoming	2		2	
Colorado	76		13	
New Mexico	19		3	
Arizona	73		14	
Utah	38		5	
Nevada	9		1	
Pacific	997	12.5	140	11.9
Alaska	1		1	
Washington	87		18	
Oregon	52		7	
California	857		114	
Hawaii	0		0	
Not reported			11	0.9
Total	7,994	100.0	1,172	100.0

(CAD) and computer graphics, management information systems (MIS), and automated materials handling (AMH) systems.[6] In this section I attempt to assess the age structure of the metalworking plants that define our universe by examining the relative sizes (and composition) of the succession of "birth" cohorts of metalworking plants. The purpose is to show that by the early 1980s a distinct cohort structure existed throughout the broad base of the U.S. metalworking sector that may influence the pace and extent of the adoption of automated machine control technologies like NC/CNC-CAM.

Information on plant age will help us understand better the range of barriers to the diffusion of technology. It is commonly suspected that greater difficulties often arise in diffusing technology through relatively old production arrangements whose physical space characteristics and supporting infrastructures are poorly matched to those of newer production and distribution arrangements.[7] Table 8 reports the birth cohort patterns among the responding plants on both chronological and historical time lines. Even though metalworking and other machine shop activities are located in some of the oldest industries in the economy, a steady process of plant cohort replacement has continuously updated the capital plant of the metalworking sector. The oldest plant in the sample began operation in 1813, but only 2.0 percent of the plants date from the nineteenth century. All through this century the growth and development of metalworking industries have been accompanied by a renewal process, in which successive plant cohorts have both expanded these industries through net plant additions and replaced older plants that have become obsolete or otherwise uncompetitive. This continuous turnover of plants and shops has made the physical capital base continually "younger."[8] Of those plants operating in 1982, only 12.9 percent began operations before the 1940s. Thus the physical capital of the plant—if not the production equipment inside—at the disposal of the small metalworking plants and shops appears to be largely of post-depression vintage. The pace of industry expansion to new plants has continued, with only a slight slackening evident during the 1970s.

Reorganization of the data by significant historical events indicates that the pace of industry expansion had accelerated dramatically for more than half a century after World War I before falling off after 1973. Although only 12.6 percent of the plants began during the interwar period 1916–1945, 24.3 percent began during 1945–1959, 39.9 percent during 1960–1973, and only 15.3 percent during 1974–1982. Let us examine the historical context for this accelerated industrial development before 1974 in more detail.

Adjustment to shifting demographic and energy conditions. During the

58

TABLE 8

DATE WHEN PRODUCTION BEGAN AT PLANT, 1800–1982

Dates	Number	Percent
Chronological time line		
1800–1849	1	0.1
1850–1874	3	0.3
1875–1899	19	1.6
1900–1909	18	1.5
1910–1919	21	1.8
1920–1929	44	3.8
1930–1939	44	3.8
1940–1949	119	10.2
1950–1959	210	17.9
1960–1969	332	28.3
1970–1979	294	25.1
1980–1982	21	1.8
Not reported	46	3.9
Total	1,172	100.1
Historical time line		
Before Civil War	2	0.2
1861–1879	8	0.7
1880–1899	13	1.1
Pre–World War I (1900–1915)	23	2.0
Interwar period (1916–1945)	140	12.6
Early post–World War II (1946–1959)	285	24.3
Era before energy adjustment (1960–1973)	468	39.9
New energy era (1974–1982)	179	15.3
Not reported	46	3.9
Total	1,172	100.0

NOTE: Percentages may not add to 100.0 because of rounding.

interwar years the United States began its emergence as the dominant military and industrial power of the world. The expansion of the broad base of plants in the metalworking sector is evidence of the kind of industrial development that was taking place to secure, preserve, and extend that dominance. During the immediate post–World War II period before 1960, the United States consolidated its industrial gains; this is clearly revealed in the dramatically increased expansion of metalworking plants. Nearly a quarter of the plants began operation during that period. But the 1960s and early 1970s witnessed the most dramatic plant expansion and development of the metalworking sector. This period was characterized by steadily rising affluence in the United States and the expanding consumer demand throughout the domestic economy that accompanied it. Population growth and household formation proceeded at ever higher rates during this period and triggered rising levels of housing consumption in mushrooming suburbs, of the automobiles needed as the separation between home and workplace increased, and of consumer durables reflecting the expanding range of home, work, and leisure activities.[9] Although the major impetus to growth was the expansion of domestic demand, inroads were also being made into foreign markets.

The effect of this expanded industrial activity was registered clearly on patterns of plant origins. Nearly four plants in ten began operation during the 1960–1973 period. The network of supplier plants expanded rapidly, and the manufacturing plants that expanded or were born often did so in locations increasingly distant from their historical locations. The transportation portion of total production costs declined for many industries and thus loosened the constraints of location between producers, suppliers, and consumers. The process was slower in the metalworking industries, however, because the shipping costs of the inputs and finished products were generally higher than those of other fabrication materials. This reinforced the necessity for the supply network of small and medium-sized plants to expand fairly close to the fabricating industries to which they supplied parts.

Finally, a defining feature of this industrial period was that energy inputs were relatively inexpensive. After 1973, however, this condition changed abruptly. The resulting shifts in the relative costs of production factors, the rising levels of imports, and the cyclical sensitivity of demand for the products of a handful of dominant industries caused widespread distress throughout metalworking industries during much of the 1970s and early 1980s. Regional economies as settings for an industry that had retained its high regional concentration now registered the effects of international markets struggling to make relatively rapid adjustments to the politically induced price boosts and production quotas engineered by the OPEC cartel beginning in late 1973. The energy price adjustments triggered by the oil embargo and the subsequent uncertainty about the availability of uninterrupted supplies had an especially unsettling effect on older basic industries, whose operations were suddenly found to be based on indulgent energy assumptions.[10] Basic and nonbasic industries alike began to respond to the prospects of high vulnerability to and growing dependence on foreign suppliers in unstable regions of the world. Furthermore, what at first looked like relatively inelastic demand over the short run accompanied by projections of rising demand over the long run conditioned the industrial activity of the nation.

The energy price adjustments induced by OPEC's intrusion in world markets, cyclical fluctuations in the economy, stagnant productivity growth, and rising foreign competition helped to dampen the pace of net plant expansion that had so characterized many metalworking industries during the preceding period. Rates of plant expansion during 1974–1982 dropped back to pre-1960 levels and thus came to reflect the shift of capital investment away from the construction of new plants and toward the acquisition of new equipment inside existing plants.[11] Certainly, the increasingly severe and extended recessions here and abroad and the simultaneous inroads made by imports from strong foreign trading partners, which created lopsided trade deficits, helped account for the dampened rate of new plant formation. By the early 1980s the remarkable strength of the dollar after the recovery reinforced the declining competitiveness of U.S. metalworking products in foreign and domestic markets.

The post-1973 era defines a period in which metalworking industries had to adjust to changes in energy resources as well as the perennial shortage of skilled labor, especially machinists. The period marks the beginning of a new surge in manufacturing automation throughout the metalworking sector that continues to the present. Amid a lingering productivity slowdown, increases were sought through modernization of plants and the production activities inside them. At one level the rising cost of energy inputs influenced plants to take steps to wring out of their production processes considerable waste of energy and materials. At another level the continuing scarcity of appropriately skilled labor in an otherwise labor-rich period helped develop a rapidly expanding market for automated machine control technologies. Nonetheless, the specific kinds of technology whose adoption is of interest in this study—expansion of NC/CNC-CAM capabilities—were not yet available at a price and in a form that most of these firms could afford or use.

Regional Patterns of Industrial Development. Inevitably, industrial change reflects product cycles at work in new and maturing industries, as well as life cycles at work in new and maturing production technologies and final products. The locational dimension of industrial change is simply a more specific reflection of the locational patterns of investment and disinvestment within and among industries. The development of the metalworking sector is interdependent with the evolution of American manufacturing itself over the past two centuries. Evidence of net plant growth radiating from the original manufacturing core regions provides the spatial features of only part of the development of the U.S. metalworking sector, however. Although the present data base allows us only to trace the net patterns of growth through plant turnover, we can use the cohort patterns revealed in the data to understand better the structural and spatial development of the metalworking sector in this century.

Table 9 displays the cohort patterns of plant origin by region. Although the industry appears to have risen rather slowly from its original Middle Atlantic and New England seedbeds in the late nineteenth century, especially after World War I, with the ascendancy of the states bordering the Great Lakes as the nation's industrial heartland, the pace of expansion in the East North Central region soon surpassed that found elsewhere. The acceleration of industrial development in the East North Central region, which continued through the 1970s, is accounted for in large part by the expansion of the automobile, consumer durable, and defense-related industries and their regional support networks of tens of thousands of parts suppliers.

The pace of expansion in the Middle Atlantic region quickened during the same period, although during the 1970s it dropped off while that of the East North Central region continued to rise. The expansion in the Far West—principally in California—occurred largely after World War II; even so, it was eclipsed in each decade by that in the East North Central and Middle Atlantic regions. Finally, the growth in the South (South Atlantic, East South Central, and West South Central) generally came later than that in the Far West. As the South gradually became the manufacturing employment center of the nation and after 1960 overtook the West as the population growth center, plant growth in many metalworking industries slowly filtered into the South. Yet, like the West immediately before it, the South has never seriously challenged the older industrial regions for dominance in the metalworking sector.

Even though the net growth of metalworking plants has spread throughout the nation in the twentieth century, new plant cohorts, just like those established in earlier decades, have been largely captured by the older industrial regions. Some 56.5 percent of the 207 new plants in the 1950s cohort, 46.6 percent of those in the 1960s cohort, and 47.2 percent of those in the 1970s cohort were located in the East North Central and Middle Atlantic regions.

Even though the sizes of new metalworking plant cohorts rose steadily in the 1940–1969 period in all regions—except for a minor reversal in the East South Central region in the 1950s—older industrial regions show evidence of an ability to compete with any other region in upgrading the physical plant capacity of metalworking production.

TABLE 9
DATE WHEN PRODUCTION BEGAN AT PLANT, BY REGION, 1800–1982

Dates	New England No.	%	Middle Atlantic No.	%	East North Central No.	%	West North Central No.	%	South Atlantic No.	%	East South Central No.	%	West South Central No.	%	Mountain No.	%	Pacific No.	%
1800–1849	—	—	1	0.5	—	—	—	—	—	—	—	—	—	—	—	—	—	—
1850–1874	1	0.9	1	0.5	—	—	—	—	1	1.4	—	—	—	—	—	—	—	—
1875–1899	3	2.8	7	3.3	2	0.5	4	3.8	2	2.9	—	—	—	—	1	2.6	—	—
1900–1909	5	4.6	3	1.4	7	1.8	—	—	1	1.4	—	—	—	—	—	—	1	0.7
1910–1919	3	2.8	5	2.4	9	2.3	—	—	1	1.4	1	4.0	1	1.4	—	—	2	1.4
1920–1929	2	1.9	8	3.8	20	5.2	4	3.8	—	—	—	—	—	—	4	10.5	3	2.1
1930–1939	6	5.6	8	3.8	19	4.9	5	4.7	—	—	—	—	2	2.9	1	2.6	2	1.4
1940–1949	7	6.5	26	12.3	44	11.3	14	13.2	6	8.7	3	12.0	8	11.4	1	2.6	8	5.7
1950–1959	15	13.9	37	17.5	80	20.6	17	16.0	8	11.6	2	8.0	13	18.6	5	13.2	30	21.4
1960–1969	36	33.3	65	30.7	87	22.4	26	24.5	28	40.6	10	40.0	20	28.6	7	18.4	47	33.6
1970–1979	26	24.1	46	21.7	91	23.5	28	26.4	18	26.1	9	36.0	21	30.0	15	39.5	36	25.7
1980–1982	—	—	2	0.9	9	2.3	4	3.8	2	2.9	—	—	1	1.4	1	2.6	2	1.4
Not reported	4	3.7	3	1.4	20	5.2	4	3.8	2	2.9	—	—	4	5.7	3	7.9	9	6.4
Total	108	100.1	212	100.2	388	100.0	106	100.0	69	99.9	25	100.0	70	100.0	38	99.9	140	99.8

NOTE: Percentages may not add to 100.0 because of rounding.

63

Even during the recession-plagued mid-1970s and early 1980s, the older regions continued to spawn new plants, fully 42.9 percent of the new plant cohort of the early 1980s being added to the capital base of the East North Central region alone. These data, then, offer little support for the common speculation that the physical capital of the older industrial regions is less susceptible to renewal and rejuvenation than that of more recently developed regions. Within this stratum of the metalworking sector, at least, even though the industry has been dispersed to all parts of the country, the regions that have dominated new plant growth since the nineteenth century continue to do so.

The implications of these patterns for the adoption and diffusion of newer technologies are considerable. There appears to be no demonstrable reason to expect that older regions are destined to lose their grip on the metalworking sector or that the process of industry renewal through successive plant cohorts must inevitably shift the critical mass of the sector from one region to another. Although certain metalworking activities have been susceptible to interregional shifts, industrial adjustment can take place by the sector's steadily relocating itself into progressively newer plants without migrating from older to newer industrial regions. This process tends to equalize the capacity of plants to adopt advanced production technologies regardless of their regional location. One would therefore expect region to wield little influence on the diffusion of technology. "The key factor is not regional location, but obsolescence." [12]

Corporate Structure of the U.S. Metalworking Industry. A focus on corporate structure draws attention to the prevalence of multiestablishment, multilocational firms in an industry. The corporate structure of U.S. manufacturing has invited special scrutiny in recent years because multiestablishment firms have demonstrated a capacity and willingness to shift production to branch or satellite plants in new locations at home and abroad. These firms can thereby adjust to shifting economic conditions through strategies of investment and disinvestment in which older industrial plants and areas may be at a distinct disadvantage. "Runaway shops," whether in the blatant form of closing in one location and reopening in another or in the more subtle form of targeting expansion to selected new locations, have given rise to considerable concern over capital mobility and the accompanying patterns of disinvestment. [13]

Although the birth of new single-site firms can lead to the same outcome and the disinvestment is just as systematic, the results are usually viewed as less calculated. Capital mobility as evidenced by branching—that is, expansion to more attractive business climates—

has been linked to the decline of older industrial regions and social and political arrangements tied to them. Special concern has been directed to the likelihood that older industrial firms are less open to rationalizing their production activities by upgrading existing plant and equipment than to shifting production to new plants in more profitable business settings. Similarly, when changes are made in place, concern has arisen that they typically cause a "deskilling" of the work force,[14] increase the proportion of part-time workers,[15] alter work rules to fragment or combine job tasks in ways that shift shop floor control more toward management, and otherwise erode the foundations of traditional occupation-income distributions established over long decades.[16] In short, industrial change is an inevitable precursor of broader social change.

The extent to which plants and shops in this stratum of the metalworking sector are tied to complex corporate structures can be expected to influence the pattern and timing of technology upgrading that takes place. Since upgrading the technology of existing plants is only one strategy of industrial renewal open to metalworking firms, whether and how it takes place are influenced by weighing its relative merits against those of alternative strategies, which may include plant relocation and other forms of capital disinvestment. Considerable concern has been expressed in recent years that the greater readiness of U.S. firms to choose cost control through capital flight than renewed investment in advanced production technologies threatens our national and older regional industrial bases.

There appears little reason to be concerned that the corporate structure of the metalworking sector can present a major barrier to technology upgrading, however. The medium-sized and smaller plant stratum of the industry is composed predominantly of free-standing establishments (table 10). Seven in ten (71.2 percent) are single-site plants, that is, are located in one place only and not tied formally to other establishments. Only 28.0 percent are part of a multiestablishment corporate structure; of this subset, one in four serves as national headquarters for a larger corporate enterprise, and three in four are regional production or product division centers.

Similarly, although this pattern varies considerably among the nine geographic regions, in each a majority of plants are independent of larger corporate structures (table 11). The most likely locations for the plants that serve as national headquarters for larger corporate structures are the Mountain (12.8 percent), East South Central (8.7 percent), Middle Atlantic (8.2 percent), East North Central (7.9 percent), and West North Central (7.7 percent) regions.

This stratum of the metalworking sector in New England, the

TABLE 10
CORPORATE STRUCTURE IN THE METALWORKING SECTOR, 1982

Corporate Structure	Number	Percent
Single-site companies	835	71.2
Multiplant companies		
National headquarters	80	6.8
Regional or product division centers	248	21.2
Not ascertained or not applicable	9	0.8
Total	1,172	100.0

Middle Atlantic, and the East North Central regions in the East and the Pacific region in the West includes relatively large portions of independent metalworking establishments. The South Atlantic, East South Central, West South Central, and West North Central regions—those peripheral to the older industrial core—are more likely to be the locations to which firms headquartered elsewhere have expanded through the years. In each of these regions, the proportion of multi-establishment enterprises that are not headquarters plants is relatively high.

Table 12 shows the extent to which the patterns of corporate structure have been influenced by the succession of plant cohorts over time. Since the 1930s approximately 70 percent of the plants in each cohort have been single-site. For both kinds of plants the largest cohort was that of the 1960s. Furthermore, the patterns of cohort size across time do not vary appreciably from one kind of plant to the other. The same pattern holds true for the historical time line.

The branching process. The imagery surrounding branch plants and the local economies tied to them that most commonly generates controversy is that of a relatively specialized and dependent production site established through relocation or secondary expansion into a location chosen for the favorable "business climate"—commonly defined in terms of low-wage labor and antiunion sentiment—it offers. Examining the subset of plants that serve as branch or satellite production platforms for larger enterprises should detect whether branching in the metalworking sector has been directed in this way. Evidence of relatively long-distance interregional branching—especially from the East North Central and Middle Atlantic regions—to locations in the relatively low-wage and nonunion South would be especially illustrative of this strategy.

TABLE 11

CORPORATE STRUCTURE IN THE METALWORKING SECTOR, BY REGION, 1982

| Region | Single-Site Plant | | Component of Multiplant Company | | | | | Total | |
			National headquarters		Regional/product division headquarters				
New England	78	(72.2)	4	(3.7)	26	(24.1)		108	(100.0)
Middle Atlantic	154	(74.0)	17	(8.2)	37	(17.8)		208	(100.0)
East North Central	269	(70.4)	30	(7.9)	83	(21.7)		382	(100.0)
West North Central	65	(62.5)	8	(7.7)	31	(29.8)		104	(100.0)
South Atlantic	47	(66.2)	3	(4.2)	21	(29.6)		71	(100.0)
East South Central	12	(52.2)	2	(8.7)	9	(39.1)		23	(100.0)
West South Central	40	(59.7)	3	(4.5)	24	(35.8)		67	(100.0)
Mountain	23	(59.0)	5	(12.8)	11	(28.2)		39	(100.0)
Pacific	117	(84.3)	4	(2.9)	17	(12.3)		138	(100.0)
Total	805	(70.6)	76	(6.7)	259	(22.7)		1,140	(100.0)

NOTE: Figures in parentheses are percentages.

67

TABLE 12
DATE WHEN PRODUCTION BEGAN AT PLANT, BY CORPORATE STRUCTURE, 1800–1982

Dates	Single-Site Plant		Component of Multiplant Company	
	Number	Percent	Number	Percent
Chronological time line				
1800–1849	1	0.1	0	0
1850–1874	3	0.4	0	0
1875–1899	10	1.2	7	2.0
1900–1909	10	1.2	8	2.3
1910–1919	15	1.8	6	1.7
1920–1929	24	3.0	18	5.2
1930–1939	32	3.9	11	3.2
1940–1949	84	10.4	33	9.6
1950–1959	144	17.8	66	19.2
1960–1969	240	29.6	87	25.3
1970–1979	209	25.8	82	23.8
1980–1982	12	1.5	9	2.6
Not reported	27	3.3	17	4.9
Total	811	100.0	344	99.8
Historical time line				
Before Civil War	2	0.2	0	0
1861–1879	5	0.6	3	0.9
1880–1899	7	0.9	4	1.2
Pre–World War I (1900–1915)	14	1.7	9	2.6
Interwar period (1916–1945)	99	12.2	45	13.1
Early post–World War II (1946–1959)	196	24.2	88	25.6
Era before energy adjustment (1960–1973)	338	41.7	122	35.5
New energy era (1974–1982)	123	15.2	56	16.3
Not reported	27	3.3	17	4.9
Total	811	100.0	344	100.1

NOTE: Percentages may not add to 100.0 because of rounding.

Not unexpectedly, the same five peripheral regions that had the lowest proportions of single-site plants have the highest proportions of branch plants other than national headquarters. Tables 13 and 14 report regional features of the branching process in the metalworking

TABLE 13

REGION OF PLANT COMPONENT OF MULTIPLANT COMPANY, 1982

Region of Parent Plant	Region of Satellite Plant									
	1	2	3	4	5	6	7	8	9	Total
1. New England	13	7	6	2	1		2		1	32
2. Middle Atlantic	3	20	15	5	5	2	8	1	2	61
3. East North Central	3	2	44	8	3	3	3	4	2	72
4. West North Central	1		3	11	1		1			17
5. South Atlantic	1	3	2		5			1	1	13
6. East South Central						3		1		4
7. West South Central	1		4	1		1	3	2	1	13
8. Mountain				1				1		2
9. Pacific	2	4	3	1	2		4	2	9	27
Europe	2		4	2	1	2	1		1	13
Total	26	36	81	31	18	11	22	12	17	254

sector. Only a quarter of the plants in the sample meet the definition of a branch or satellite plant. In general, no substantial targeting of the South Atlantic and the East and West South Central regions by headquarters plants in the East North Central and Middle Atlantic regions appears to have taken place. Only 24.6 percent of the branch plants of firms headquartered in the Middle Atlantic region and only 12.5 percent of the branch plants of firms headquartered in the East North Central region were located in one of the three southern regions (table 13). For headquarters firms in the dominant East North Central region, any branching was as likely to be directed to the West North Central region as to the three southern regions combined.

Overall, the most prevalent pattern was for branch plants to be located in the same region as the parent plant. Parent plants are most likely to be located in the older industrial Middle Atlantic and East North Central regions (table 14). Moreover, while 32.0 percent of all branch plants are located in the East North Central, the proportions of branch plants throughout the older industrial regions are considerably higher than they are outside them. This pattern, together with the data in table 13, suggests that the branching that has occurred has been over relatively short distances. Although metalworking plants in

TABLE 14

BRANCHING IN THE U.S. METALWORKING SECTOR, 1982

Region	Location of Branch Plants		Location of U.S. Parent Plants		Intraregional Branching		Interregional Branching		Percentage of Intraregional Branches	Percentage of Interregional Branches
	No.	%	No.	%	No.	%	No.	%		
New England	24	10.0	32	13.3	13	11.9	19	14.4	40.6	59.4
Middle Atlantic	36	14.9	61	25.3	20	18.3	41	31.1	32.8	67.2
East North Central	77	32.0	72	29.9	44	40.4	28	21.2	61.1	38.9
West North Central	29	12.0	17	7.1	11	10.1	6	4.5	64.7	35.3
South Atlantic	17	7.1	13	5.4	5	4.6	8	6.1	38.5	61.5
East South Central	9	3.7	4	1.7	3	2.8	1	0.8	75.0	25.0
West South Central	21	8.7	13	5.4	3	2.8	10	7.6	23.1	76.9
Mountain	12	5.0	2	0.8	1	0.9	1	0.8	50.0	50.0
Pacific	16	6.6	27	11.2	9	8.3	18	13.6	33.3	66.7
Total	241	100.0	241	100.1	109	100.1	132	100.1	45.2	54.8

NOTE: Percentages may not add to 100.0 because of rounding.

the South are somewhat more likely to be branch plants, branch plants are more likely to be located in the older industrial regions alongside their parent plants. Thus evidence of a "southern strategy," whereby plants in older, high-cost industrial areas have systematically sought to locate branches in the low-cost South, is slim at best.[17] Of all the branching in this data set, 45.2 percent was confined within regions, and 54.8 percent was interregional. Intraregional branching has been most common in the East North Central region. Furthermore, of all the interregional branching that has occurred, 12.9 percent has involved headquarters plants in the East North Central and Middle Atlantic regions placing branch plants in each other's region.

Table 14 also reports the proportion of the branches of parent plants in each region that are located in that region. Of the seventy-two parent plants located in the East North Central region, for example, forty-four (61.1 percent) of the branches are in the same region. The proportion is highest in the East North Central, the West North Central, the East South Central, and the Mountain regions.

In examining where firms headquartered in the South have branched, an even more interesting finding emerges. A third (36.7 percent) of the branches of metalworking firms headquartered in one of the three southern regions are located in the East North Central, the Middle Atlantic, and the New England regions, and an equal proportion are located in the same region as the parent. Once again, not region per se but the overall geographical structure of the metalworking sector appears to influence the branching process. Most parent plants and their branches are concentrated in the older industrial regions, and headquarters plants in the South and even the West are as likely to have branch plants in the older industrial regions as in the region of the parent plant.

In conclusion, although the branching process varies by region, tables 11 through 14 generally support the conclusions that much branching—which is relatively common throughout the metalworking industry—occurs in the region of the parent plant. The branching of parent plants in older regions to plants in either the South or the West is relatively uncommon. Accordingly, this data set offers little evidence of a "southern strategy" whereby northern plants systematically direct mobile capital to distant locations. The high concentration of the sector in older industrial regions does not appear to be in danger of being significantly diminished by the branching that has occurred or will probably occur in the foreseeable future.

Unionization Trends and Shifting Labor Environments. Unions have commonly been identified as impediments to the flexibility needed by an industry as it attempts to adjust to shifting economic conditions. From this perspective unions are often viewed as tending to preserve increasingly outmoded, inappropriate, and unresponsive labor environments within plants and inflexibility within an industry. From another perspective, however, unions are viewed as preventing necessary adjustments from imposing unnecessarily heavy burdens on workers.[18] The conflict in perspectives is too complex to be resolved by a single data set, but how an entire industry can gradually migrate from one kind of labor environment to another can be illustrated by the data reported here.

The union status of production workers is commonly believed to influence patterns of employment change in many industrial sectors and settings. To the extent that union status is tied to wage and benefit packages and formally defined work rules that govern those who are employed and to relatively generous supplemental unemployment benefits for those who are unemployed, the resulting "social wage" can greatly increase labor costs and thus intensify the employment contraction that may be sweeping through an industry in response either to short-term cyclical pressures or to longer-term erosion of markets resulting from rising imports or declining domestic demand. This more chronic effect can be completely independent of any employment contraction caused by the adoption of automated production technologies.

To what extent does unionization influence whether, why, or when metalworking firms adopt automated machine control technologies? This chapter explores the role of unionization in industrial adjustment at two levels. The distribution of right-to-work provisions among states establishes an initial influence on the distribution of unionized plants. Within states at individual plants and shops, unionization is then viewed as a defining feature of specific production settings. The data presented below suggest that although the influence of organized labor in medium-sized and small metalworking plants and shops has never been great, a prior long-term industrial adjustment involving a transit between labor environments has been under way. Major portions of the metalworking sector have gradually migrated out of unionized labor settings at the plant level without migrating to states that have used right-to-work legal provisions to discourage inroads by unions. The result has been a further erosion of the role of unions throughout the metalworking sector.

Only 22.5 percent of the metalworking plants in the sample have unionized production workers (see table 15). This national percentage is exceeded in only three areas—the East South Central, the Middle Atlantic, and the East North Central regions.

TABLE 15
Union Status of Production Workers, by Region, 1982

Region	Union		Nonunion		Total	
	No.	%	No.	%	No.	%
New England	19	17.8	88	82.2	107	100.0
Middle Atlantic	62	30.1	144	69.9	206	100.0
East North Central	101	26.9	274	73.1	375	100.0
West North Central	20	19.8	81	80.2	101	100.0
South Atlantic	6	8.8	62	91.2	68	100.0
East South Central	8	32.0	17	68.0	25	100.0
West South Central	12	17.4	57	82.6	69	100.0
Mountain	4	10.5	34	89.5	38	100.0
Pacific	21	15.4	115	84.6	136	100.0
Total	253	22.5	872	77.5	1,125	100.0

73

The role of right-to-work legislation. Before exploring in greater detail the influence on industrial adjustment of unionization understood as a characteristic of the work setting inside metalworking plants and shops, let us consider the larger context that defines the "business climate" and that may offer a relatively accommodative or unsupportive environment for unionized labor settings on a larger scale.

To what extent is this pattern influenced by states' right-to-work provisions? Such provisions exist in twenty states. Since their influence is often thought to inhibit unionization, we might expect to see some correlation between rates of unionization in plants in the metalworking sector and the existence of right-to-work provisions. Unionization rates are indeed significantly related to right-to-work status (table 16). Only 16.8 percent of all metalworking plants are located in right-to-work states, and only 15.3 percent of the plants in those states are unionized. The vast majority of plants (76.0 percent) in non-right-to-work states are also nonunion, however.

Today a semicircle of states with right-to-work provisions girds the industrial heartland and serves as a buffer between it and the states to the west. Table 17 displays the succession of plant cohorts for states with and without right-to-work provisions. The chronological time line reveals a long-term decline in the share of each plant cohort located outside the band of right-to-work states. The share declined from 95.3 percent during the 1920s to 75.9 percent during the 1970s and to 71.4 percent during 1980–1982. The historical time line reveals a somewhat different pattern, however. Although nearly all (95.7 percent) of the plants begun during 1900–1915 and still operating in the early 1980s were located in non-right-to-work states, that proportion dropped to 86.6 percent for plants begun just after World War II (1946–1959), to 81.8 percent for plants begun during 1960–1973, and to 72.6 percent for plants begun after 1973. Therefore, the vast majority of existing metalworking plants are located in non-right-to-work states regardless of when they were founded. The regional concentration of plants and shops has not been influenced appreciably as the larger industry has adjusted to the enactment of right-to-work legislation by states.

The role of unionized production workers. Has a relatively low proportion of unionized metalworking plants always characterized metalworking, or has a retreat from unionized production settings characterized the continued evolution of the sector? In other words, is there any evidence that unionization has had anything but the relatively modest influence that it now has in this segment of the U.S. metalworking sector? Table 18 reports the distribution of union and nonunion plants among the succession of plant cohorts. The cohort sizes of

74

TABLE 16
UNION STATUS OF PRODUCTION WORKERS,
BY RIGHT-TO-WORK STATUS OF STATES, 1982

Union Status	Right-to-Work State		Non-Right-to-Work State		Total	
	No.	Percent	No.	Percent	No.	Percent
Union	29	15.3	226	24.0	255	22.5
Nonunion	161	84.7	717	76.0	878	77.5
Total	190	16.8	943	83.2	1,133	100.0

NOTE: χ^2 = 6.86; degrees of freedom = 1; significant at .01 level.

both union and nonunion plants have generally increased all through the twentieth century; however, although the size of successive cohorts of union plants continued to increase through the 1960s before a major decline during the 1970s, those of nonunion plants continued to increase through the 1960s without a significant decline during the 1970s. Among plants existing in the early 1980s, the majority of the oldest ones (59.6 percent)—those begun in the 1920s or earlier—were unionized. Since then, however, nonunion plants have captured larger majority shares of each succeeding cohort. The nonunion share has risen steadily from 60.0 percent for the 1930s cohort, 75.0 percent for the 1950s cohort, and 92.0 percent for the 1970s cohort to 100.0 percent for the 1980–1982 cohort.

This same basic trend is revealed even more dramatically in historical terms. Nearly one-fourth (24.7 percent) of all the union plants operating in the early 1980s were begun between the two world wars (1916–1945). Roughly equal portions of union plants were begun in the early post–World War II era (1946–1959) and between 1960 and 1973 (27.5 percent). Only 4.3 percent of all union plants have begun since 1974. By comparison, only 8.9 percent of all nonunion plants began between the world wars and 23.9 percent in the immediate post–World War II era (1946–1959). More than four in ten (43.8 percent) began operations between 1960 and 1973 and another 18.8 percent since 1974. Although 44.7 percent of the plants begun in the interwar period are unionized, 75.0 percent of those begun during 1946–1959, 84.6 percent of those begun between 1960 and 1973, and 93.8 percent of those begun after 1973 are nonunion.

Clearly, most of the more recent metalworking plants have a nonunion production environment. The broad base of the metalworking

TABLE 17
DATE WHEN PRODUCTION BEGAN AT PLANT,
BY RIGHT-TO-WORK STATUS OF STATES, 1800–1982

Dates	Right-to-Work States		Non-Right-to-Work States	
	Number	Percent	Number	Percent
Chronological time line				
1800–1849	—	—	1	0.1
1850–1874	—	—	3	0.3
1875–1899	4	2.1	14	1.4
1900–1909	1	0.5	17	1.7
1910–1919	2	1.0	19	1.9
1920–1929	2	1.0	41	4.2
1930–1939	3	1.5	40	4.1
1940–1949	21	10.8	96	9.8
1950–1959	22	11.3	188	19.3
1960–1969	57	29.2	272	27.9
1970–1979	70	35.9	221	22.6
1980–1982	6	3.1	15	1.5
Not reported	7	3.6	49	5.0
Total	195	100.0	976	99.8
Historical time line				
Before Civil War	—	—	2	0.2
1861–1879	—	—	8	0.8
1880–1899	4	2.1	8	0.8
Pre–World War I (1900–1915)	1	0.5	22	2.3
Interwar period (1916–1945)	12	6.2	133	13.6
Early post–World War II (1946–1959)	38	19.5	246	25.2
Era before energy adjustment (1960–1973)	84	43.1	378	38.7
New energy era (1974–1982)	49	25.1	130	13.3
Not reported	7	3.6	49	5.0
Total	195	100.1	976	99.9

NOTE: Percentages may not add to 100.0 because of rounding.

sector has thus been gradually jettisoning the unionized production arrangements that more fully characterized it earlier in the century.[19]

Unionization and corporate structure. To what extent has the corporate structure of this segment of the metalworking sector influenced the gradual retreat from unionized labor environments on the shop

TABLE 18
DATE WHEN PRODUCTION BEGAN AT PLANT, BY UNION STATUS OF
PRODUCTION WORKERS, 1800–1982

Dates	Plants with Unionized Production Workers		Nonunion Plants	
	Number	Percent	Number	Percent
Chronological time line				
1800–1849	1	0.4	—	—
1850–1874	2	0.8	1	0.1
1875–1899	13	5.1	5	0.6
1900–1909	14	5.5	4	0.5
1910–1919	8	3.1	13	1.5
1920–1929	24	9.4	19	2.2
1930–1939	16	6.3	24	2.7
1940–1949	35	13.7	79	9.0
1950–1959	52	20.4	156	17.8
1960–1969	58	22.7	267	30.4
1970–1979	23	9.0	264	30.1
1980–1982	—	—	19	2.2
Not reported	9	3.5	27	3.1
Total	255	99.9	878	100.2
Historical time line				
Before Civil War	1	0.4	1	0.1
1861–1879	7	2.7	1	0.1
1880–1899	8	3.1	4	0.5
Pre–World War I (1900–1915)	16	6.3	7	0.8
Interwar period (1916–1945)	63	24.7	78	8.9
Early post–World War II (1946–1959)	70	27.5	210	23.9
Era before energy adjustment (1960–1973)	70	27.5	385	43.8
New energy era (1974–1982)	11	4.3	165	18.8
Not reported	9	3.5	27	3.1
Total	255	100.0	878	100.0

NOTE: Percentages may not add to 100.0 because of rounding.

floor? Table 19 shows a significant relation between union status and whether a plant is independent or part of a multiplant corporate structure. Plants tied to more complex corporate structures are more likely than independent plants to have unionized production settings. While the vast majority (85.2 percent) of single-site plants are nonun-

TABLE 19
UNION STATUS OF PRODUCTION WORKERS, BY PLANT TYPE, 1982

Union Status	Single-Site Plant		Component of Multiplant Company		Total	
	No.	Percent	No.	Percent	No.	Percent
Union	117	14.8	136	40.4	253	22.4
Nonunion	675	85.2	201	59.6	876	77.6
Total	792	70.2	337	29.8	1,129	100.0

NOTE: $\chi^2 = 89.0$; degrees of freedom = 1; significant at .001 level.

ion, only 59.6 percent of the plants in multiestablishment firms are. Therefore, the data suggest that the growth of single-site plants—rather than processes such as branching, in which lines of control over shop floor activities are extended to headquarters plants—has spurred the growth of nonunion production settings in and beyond the buffer zone of non-right-to-work states. In other words, the shift of growth into new, independent plants has been a vehicle through which this segment of the metalworking sector has gradually abandoned unionized production settings.

In conclusion, despite the spread of metalworking plants through all parts of the country, the high degree of geographical concentration that has always defined this industry endures. Evidence of industrial adjustment includes a steady replacement and upgrading of physical plant through cohort succession and a branching process that has largely been contained within the regions of origin. Through the differential growth rates of unionized and nonunionized plant cohorts, there has been a steady filtering out of unionized production settings. These industrial adjustments are an important part of the context within which to interpret the patterns of diffusion and implementation of advanced production technology taken up in the next chapter.

Notes

1. The U.S. Census Bureau classifies states into nine census divisions, which can be grouped into four regions. The term "region" is used in this study to refer to multistate aggregations of both kinds.

2. T. J. Noyelle and T. M. Stanback, Jr., *The Economic Transformation of American Cities* (Totowa, N.J.: Rowman and Allanheld, 1984).

3. For a discussion of the way in which the concentration and centralization

of corporate administrative control over production coexist with the increasing deconcentration of standardized production facilities, see B. Bluestone and B. Harrison, *The Deindustrialization of America: Plant Closings, Community Abandonment, and the Dismantling of Basic Industry* (New York: Basic Books, 1982), pp. 118ff. See also Noyelle and Stanback, *Economic Transformation*; and R. Cohen, "The Internationalization of Capital and U.S. Cities" (Ph.D. dissertation, New School for Social Research, 1979).

4. See also table A-1 in the appendix. On average, 63.8 percent of all plants with automated machine control are small, although states and regions vary, with the South lagging somewhat.

5. For a corresponding overview of the U.S. machine tool industry, see National Academy of Engineering, *The Competitive Status of the U.S. Machine Tool Industry* (Washington, D.C.: National Academy Press, 1983), pp. 16ff.

6. For more complete descriptions of these forms of computerized manufacturing automation, see Office of Technology Assessment, *Computerized Manufacturing Automation: Employment, Education, and the Workplace* (Washington, D.C., 1984), pp. 35ff.

7. See D. W. Austin and J. E. Beazley, "Struggling Industries in Nation's Heartland Speed Up Automation," *Wall Street Journal*, April 4, 1983.

8. The cohort analytic format used here can offer no estimates of the absolute number of metalworking plants that began business during these time periods. It picks up net replacements for a single point in time rather than total business formations and failures over time. Business formations that ended in failure or "death" before mid-1982 are missed entirely. It is also likely that a slightly larger number of firms have been in operation continuously for longer periods than the cohort structure indicates. Firms that moved from older to newer plant and shop settings after their original founding are captured in the data in the plant cohort of their most recent location. Therefore, plant age is not identical with the date of business formation in all instances. Nonetheless, given the absence of better data sources and the definition of the original sampling frame, these constraints are expected to have no material effect on the interpretation of trends. Although these omissions are systematic and may somewhat distort our understanding of the historical patterns of industrial expansion, contraction, and overall development within industries, the cohort analytic approach is well suited to exploration of the ways in which an industry adjusts to changing circumstances through plant cohort replacement.

9. On this latter point, see G. Sternlieb, J. W. Hughes, and C. O. Hughes, *Demographic Trends and Economic Reality: Planning and Marketing in the '80s* (New Brunswick, N.J.: Center for Urban Policy Research, Rutgers University, 1982).

10. The increased energy efficiency of the U.S. economy is documented in the *1985 U.S. Industrial Outlook* (appendix), which reports that the 22 percent increase in real GNP from 1973 to 1983 was accompanied by a 23 percent decline in the energy required to produce a dollar of GNP. For durable goods manufacturing industries, the 57.5 percent increase in total energy consumption during 1958–1973 was followed by a 19.8 percent decline during 1973–1981.

11. The upgrading of physical capital through plant replacement in this segment of the metalworking sector is notable in light of related capital-labor trends in recent decades. All through the post–World War II period, capital-labor ratios have risen in tandem with measures of labor productivity in the private sector; but separate capital-labor ratios tied to two forms of capital—plant and equipment—have differed significantly throughout this period, especially during the 1970s. According to data provided by the Bureau of Labor Statistics and the Bureau of Economic Analysis of the Department of Commerce, the ratio of capital equipment to labor rose at an average annual rate of 3.7 percent between 1947 and 1980, while output per hour rose at 2.6 percent and the ratio of capital structures to labor rose at only 2.1 percent. After 1973 the gap between the separate capital-labor ratios widened, and the rates were 2.3 percent, 0.6 percent, and −0.2 percent, respectively. See Bureau of Labor Statistics, *Productivity Chartbook* (Washington, D.C. 1981). See also related discussion and data in *U.S. News and World Report*, October 8, 1984.

12. Quotation from R. Hanson, ed., *Rethinking Urban Policy: Urban Development in an Advanced Economy* (Washington, D.C.: National Academy Press, 1983), p. 35. J. Rees et al., after discovering regional variation in adoption rates for seven advanced production technologies in six four-digit industries, reached essentially the same conclusion: "It is the older industrial regions of the North Central and Northeastern parts of the Manufacturing Belt that display the highest propensity to use new production technology." J. Rees, R. Briggs, and R. Oakey, "The Adoption of New Technology in the American Machinery Industry," *Regional Studies*, vol. 18, no. 6 (1984), p. 497.

13. The best-developed statement of heightened anxiety over the undesirable consequences of unchecked capital mobility is in Bluestone and Harrison, *Deindustrialization of America*.

14. For an early statement of the concern about deskilling, see H. Braverman, *Labor and Monopoly Capital* (New York: Monthly Review Press, 1974).

15. See T. M. Stanback, Jr., P. J. Bearse, T. J. Noyelle, and R. A. Karasek, *Services: The New Economy* (Totowa, N.J.: Rowman and Allanheld, 1981), pp. 82ff.

16. See L. Thurow, "The Disappearance of the Middle Class," *New York Times*, February 5, 1984. Among the stronger challenges to this view are R. Z. Lawrence, "Sectoral Shifts and the Size of the Middle Class," *Brookings Review*, vol. 3, no. 1 (Fall 1984), pp. 3–11; and R. J. Samuelson, "Middle-Class Media Myth," *National Journal*, December 31, 1983, pp. 2673–78.

17. See "Emerson Electric: High Profits from Low Tech," *Business Week*, April 4 ,1983.

18. R. B. Freeman and J. L. Medoff, *What Do Unions Do?* (New York: Basic Books, 1984).

19. This is not to suggest that the bulk of employment or value added is derived from nonunion plants. Rather, the point here is that the climate for business formation throughout the metalworking sector has steadily screened out unionized operations.

5

The Adoption and Diffusion of Automated Production Technology

The U.S. metalworking industry complex has been adapting to an ever-changing economic environment in a variety of ways all through this century. Shifts into new physical plants and labor environments have continuously restructured the industry in several respects, if not necessarily sustained its competitiveness at the same time. Much of the resulting adjustment of the industry can be traced to the replacement of older plants and the physical production settings they offer by newer ones. As increasingly large cohorts of more recent plants have been added, the relative influence of earlier plants has been diminished. Simultaneously, the larger industry has gradually abandoned older unionized labor environments without substantially abandoning its historical regional locations. The impetus and anticipated effects of other forms of industrial adjustment, including the adoption of automated manufacturing systems, must be understood against the backdrop of such larger industrial shifts.

In this chapter I turn to the upgrading of production technology in the metalworking sector. The productivity and competitiveness of entire industries may increase steadily through plant turnover and cohort successions of the kind I have reported, but here I examine patterns of adoption of automated machine control technologies resulting from the capital investment strategies of existing plants rather than the more indirect upgrading brought about by the changing composition of plants. Ultimately, this chapter and the one that follows seek to explore the dynamics of technological innovation in an industrial setting in ways that contribute to the analysis of public policy options to guide the industrial renewal of older basic industries.[1]

To what extent have plants and shops in this segment of the metalworking sector pursued an explicit strategy of upgrading their production technologies on the shop floor? In particular, what can we learn about the adoption of specific NC/CNC-CAM machine control technologies and their diffusion throughout the sector? In addition to the patterns of this diffusion, what have been the incentives and motives for it? Have plant managers adopted automated machine control systems primarily to increase their productivity and thereby to establish a more secure competitive position, or have they used technology upgrading to wrest further control over the shop floor from machinists and other skilled workers, who by virtue of their skills and experience in conventional machine tool control have traditionally heavily influenced the ordering and pace of production? Has automated machine tool control finally begun to filter into smaller machine shops as simply one more strategy to compensate for the perennial shortage of skilled workers or because it promises to hasten the transformation of the labor environment and thereby supplements the gradual migration of the industry out of unionized work settings? Or has the evolution of NC/CNC technologies finally made machine tool control systems accessible to plants and shops in this segment of the sector by making them of appropriate scale and price to be economically feasible? If the momentum of technology development alone is found to be important, can we discover whether the rise of new potential markets triggered it or whether the newly tailored machine control technologies created their own markets? These and related questions are addressed below.

Recent Evidence of Technology Diffusion

Innovation is a relative term in the sense that what is new is the role of a particular technology in a capacity or setting in which it has not been present before. The extent to which advanced technologies are introduced into materials requirement planning, machine tool control, shop floor data collection and analysis, computer graphics and design simulation, computer simulation of manufacturing operations, and work measurement and inspection can serve as a useful indicator of industrial renewal through technology upgrading. It has been estimated that in 1968 only 0.5 percent of the machine tools in the United States were numerically controlled. That share increased only to 2.0 percent by the 1976–1978 period and only to 4.7 percent by 1983.[2] In addition to the accelerating pace of diffusion, what evidence do we have that metalworking plants and shops have begun either to adopt or to make plans to adopt in the near future advanced machine tool control technologies and assimilate them into their operations?

Mansfield and others, in a study that traced the spread of numerically controlled machine tools through ten manufacturing industries, provided a model for several diffusion studies that have followed, including this one.[3] Invariably, different technologies diffuse throughout different industries at different paces.[4] A more recent study of the use of computers in manufacturing found that the pace of planned adoptions of computer-assisted design (CAD) and computer-assisted manufacturing (CAM) systems was expected to be quite high in the future.[5] Moreover, among plants that had not yet established computerized manufacturing data bases but planned to do so, 98.8 percent planned to do so by 1984. The same study reported that current and planned use of CAD systems was closely related to the employment size of the plant. Upgrading technology was most commonly done by relatively large plants.[6] Actual and planned CAD use was greatest (60.9 percent) among plants employing more than 1,000 and least (7.3 percent) among plants employing 50 to 99. The same pattern held for actual and planned use of CAM systems, with the greatest use (64.5 percent) in the largest plants and the least use (27.2 percent) among the smaller plants. Among plants with both CAD and CAM systems, only 6.0 percent reported that they had integrated those systems. Integration was much more common among the larger plants (15.9 percent, with another 55.8 percent planning to integrate) than among the smaller plants (2.1 percent, with only 14.7 percent planning to integrate).

Thus not only is there evidence of considerable interest and actual activity in technology upgrading throughout this older manufacturing industry group, but the larger plants are leading the way. How are we to interpret the apparent importance of plant size for technology upgrading? Might these trends suggest that automated machine control technologies face special barriers when filtering down into and through the broad base of the metalworking sector? If that is so, is it because of the shortsightedness of plant managers in these strata or their inability to appreciate the benefits of new technologies? Or do these trends indicate that smaller plants and shops are making rational assessments of the potential effects of the new technologies on their operations? Perhaps the volume, scale, and composition of workloads in these plants are such that the benefits of automation are often illusory at best. If so, insufficient evidence of economic feasibility is a more important barrier than lack of managerial sophistication to the broader diffusion of NC/CNC-CAM throughout the sector.

If plant size continues to be an important factor in decisions to adopt new technology, does this constitute evidence that big business possesses relatively greater capacities for adjustment to changing economic circumstances and that advanced production technologies will polarize individual industries into competitive large and uncompetitive smaller operations? Probably not. The strategy of this study has been to begin by granting the existence of considerable evidence—both logical and empirical—that much in the way of industrial renewal and rejuvenation is likely to come from our older industrial sectors. It is accepted as true in this study that the larger companies and plants in these industries can be expected to lead the way. The evolution of automated machine tool control technologies is such, however, that the barrier of plant size understood as a proxy for the compelling economies of scale that have traditionally accelerated adoptions among larger plants can be expected to fall. As a result, the evolution of NC/CNC technology in recent years has begun to produce systems whose scale and price permit them to pass the test of economic feasibility among ever smaller metalworking plants.

How complete and extensive can the resulting upgrading be expected to be? In what ways will technology upgrading proceed in the thousands of smaller plants and shops of the durable-goods-producing sector? Is the unexpected vitality in older industrial sectors confined mainly to larger businesses, or can the small and often independent establishments in older metalworking industries exhibit a vitality similar to that so often associated with small business in newer industrial sectors? To shed light on these and related questions is a primary objective of the analyses to which I now turn.

Cohort Patterns of Technology Adoption

I begin by examining once again cohort patterns among plants. In the previous chapter, we discovered that the obsolescence of an industry's aggregate capital plant is continuously being diminished by the infusion of new plants and shops that replace and augment existing ones. If old industries do not remain confined to old plants, the regions in which they may remain concentrated for a century and more are inappropriately viewed as old. A regional comparative advantage may endure without an aging of region and industry in any meaningful sense, but can the age of a plant somehow function as a barrier to industrial adjustment through technology upgrading inside?

At any one time plants and shops in any industry can vary considerably in age. Although plant age should not be equated with the age of production capital or its arrangement inside, older plants are presumed less likely to be organized around newer technologies than plants that have been established more recently. For reasons related to custom, habit, or the inertia of older plants' operations, is it reasonable to expect that plant age itself may be a barrier to widespread adoption of new production technologies? In other words, do plant age and its correlates indicate an obsolescence that influences the diffusion of new production technologies?

Table 20 reports the timing of NC/CNC technology adoption by plant cohort. The table traces the diffusion of NC/CNC capabilities through twenty-one plant cohorts covering nearly two centuries. Since versions of numerically controlled machine tools have been available commercially since the early 1950s, we can monitor their diffusion through seven historical periods and thus can attempt to disentangle the influence of plant age from influences present during the period in which plants actually adopted NC/CNC systems. If obsolescence as a correlate of plant age is an effective barrier to upgrading technology from conventional to some form of automated machine tool control, we would expect to see technology diffusion limited to a subset of relatively recent plant cohorts. If the patterns of diffusion are spread broadly across cohorts, however, the influence of plant age or cohort membership appears to be less compelling than broader economic and technological influences.

Different periods since the early 1950s have been defined by different combinations or stages of NC/CNC technological development and economic conditions that have been more or less conducive to the adoption of new technology. The right combination of these factors at the right moment in economic history should trigger the diffusion process by prompting adoption decisions by plants and shops. Evidence of the compelling influence of the right combination of these factors would be revealed in a pattern of widespread adoption within a relatively narrow period. That is, the greater importance of period influences over cohort influences would be revealed by diffusion sequences beginning across a wide range of plant cohorts largely during the same period.

The patterns revealed in table 20 are remarkably clear. The diffusion process—among plants in which it took place at all—appears to have commenced during the 1960–1964 period. This suggests that the lag between innovation and diffusion was somewhat less than a decade.[7] Of all the plants in our sample for which data are available, fewer than 1.0 percent adopted NC technology in the same decade in which it was developed. Although the initial trickle of adoption of automated machine control began in the early 1960s, the pace was slow. Cohorts of plants established before 1960 typically adopted progressively higher proportions of new technology over the next twenty years. Not until the late 1970s—a full quarter-century after the initial innovation wave—did adoption rates peak. For plants established since the early 1960s, the pattern is the same. The adoption of automated machine tool control did not peak until the 1975–1979 period,

TABLE 20

Date of Adoption of NC/CNC Technology, by Year Plant Began, 1800–1982

Year Plant Began	Age of Plant (years)	Before 1950	1950–1954	1955–1959	1960–1964	1965–1969	1970–1974	1975–1979	1980–1982	Total
1800–1824	158–182	—	—	—	1 (100.0)	—	—	—	—	1 (100.0)
1825–1849	133–157	—	—	—	—	—	—	—	—	—
1850–1874	108–132	—	—	—	—	2 (67.0)	—	1 (33.0)	—	3 (100.0)
1875–1899	83–107	—	—	—	1 (5.9)	11 (64.7)	1 (5.9)	2 (11.8)	2 (11.8)	17 (100.1)
1900–1904	78–82	—	—	—	—	3 (25.0)	5 (41.7)	3 (25.0)	1 (8.3)	12 (100.0)
1905–1909	73–77	—	—	—	—	1 (20.0)	1 (20.0)	2 (40.0)	1 (20.0)	5 (100.0)
1910–1914	68–72	—	—	—	—	—	1 (20.0)	4 (80.0)	—	5 (100.0)
1915–1919	63–67	—	—	—	2 (13.3)	3 (20.0)	4 (26.7)	4 (26.7)	2 (13.3)	15 (100.0)
1920–1924	58–62	1 (4.5)	—	—	3 (13.6)	3 (13.6)	7 (31.8)	8 (36.4)	—	22 (99.9)

Year	Age									Total
1925–1929	53–57	—	—	—	2 (10.0)	5 (25.0)	3 (15.0)	9 (45.0)	1 (5.0)	20 (100.0)
1930–1934	48–52	—	—	—	3 (18.8)	2 (12.5)	3 (18.8)	5 (31.3)	3 (18.8)	16 (100.2)
1935–1939	43–47	—	—	—	1 (4.0)	7 (28.0)	3 (12.0)	13 (52.0)	1 (4.0)	25 (100.0)
1940–1944	38–42	—	—	—	2 (6.9)	6 (20.7)	7 (24.1)	13 (44.8)	1 (3.4)	29 (99.9)
1945–1949	33–37	—	—	—	3 (3.6)	14 (16.7)	20 (23.8)	41 (48.8)	6 (7.1)	84 (100.0)
1950–1954	28–32	—	1 (1.0)	—	8 (8.2)	21 (21.6)	26 (26.8)	34 (35.1)	7 (7.2)	97 (99.9)
1955–1959	23–27	—	—	2 (2.3)	5 (5.8)	21 (24.4)	20 (23.3)	27 (31.4)	11 (12.8)	86 (100.0)
1960–1964	18–22	—	—	—	10 (8.1)	22 (17.9)	29 (23.6)	47 (38.2)	15 (12.2)	123 (100.0)
1965–1969	13–17	—	—	—	—	47 (23.7)	49 (24.7)	80 (40.4)	22 (11.1)	198 (99.9)
1970–1974	8–12	—	—	—	—	—	53 (35.3)	70 (46.7)	27 (18.0)	150 (100.0)
1975–1979	3–7	—	—	—	—	—	—	77 (77.0)	23 (23.0)	100 (100.0)

(Table continues)

87

TABLE 20 (continued)

Year Plant Began	Age of Plant	Before 1950	1950–1954	1955–1959	1960–1964	1965–1969	1970–1974	1975–1979	1980–1982	Total
1980–1982	0–2	—	—	—	—	—	—	—	7 (100.0)	7 (100.0)
Total		1 (0.1)	1 (0.1)	2 (0.2)	41 (4.0)	168 (16.6)	232 (22.9)	440 (43.3)	130 (12.8)	1,015 (100.0)

NOTE: Figures in parentheses are percentages. Percentages may not add to 100.0 because of rounding.

although progressively larger proportions of each successive plant cohort since the early 1960s began operations with at least minimal NC/CNC capabilities in place. The availability of CNC helped make machine tool automation increasingly accessible to small plants and shops.[8] Moreover, this diffusion was not facilitated by extraordinary technology-promoting policy measures of the kind widely discussed during the business cycle downturns of the early 1980s. A constellation of factors facilitated technology upgrading and the capital investment it symbolized even before the extended downturns began.

The evidence suggests that neither the the availability (since the 1950s) of basic forms of numerical control technology nor the age of the physical plant into whose operations it must be integrated is as important as the market conditions that define industrial competitiveness at a particular time. The availability of smaller and less expensive turnkey systems (that is, hardware-software combinations tailored to specific industry settings) to smaller companies and the need to improve productivity to remain competitive may be the principal inducements for the adoption by metalworking firms of relatively advanced production technologies like NC and CNC.

The relative importance of the diffusion of new production technology, as opposed to its mere existence, is revealed in these data.[9] An evolving lineage of machine control technologies existed a full quarter-century before any adoption took place throughout the broad base of the metalworking sector. Until the right combination of scale and price characteristics tailored to the dominant features of plants in this stratum was available, machine tool control technologies remained both inaccessible and inappropriate. With regard to plant age and obsolescence, then, economic conditions and the life stages of the new technology have influenced the NC/CNC diffusion process more clearly than cohort membership and its correlates. Old metalworking plants faced with the right incentives and accessible technological alternatives appear to have the capacity to help reposition an entire industry and improve its competitiveness by upgrading technology. The extent to which this capacity is exercised, however, is another matter.

To the Plant versus through the Plant:
Diffusion versus Penetration and Saturation

Studies of technology diffusion commonly trace the spread of innovations from plant to plant through an industry or across an economic landscape. Has a particular technology been adopted more quickly in the metal castings industry or in the electrical appliance industry, in the automobile or the aircraft industry? What characteristics of plants

89

and their operations channel diffusion, accelerate or impede it, or exclude entire segments of an industry from the process altogether?[10] Any inquiry about the role of technology adoption in the rationalization of industrial activities (production, shipping, materials handling, marketing, and sales) is only partially informed by knowing whether or when a particular plant adopted a particular technology. Even more important is knowledge about the extent of that adoption and the degree to which new technologies have come to dominate what actually goes on inside the plant.

New technology getting *to* the plant is arguably less important than technology wending its way *through* the plant and leaving its imprint on the full range of industrial production activities taking place inside. The case of numerical control in U.S. metalworking plants offers a good example. We know that less than 4 percent of all machine tools in the United States were numerically controlled by 1983, even though the technology had diffused to plants in all regions.[11] If the purpose of upgrading technology is to enhance a plant's chances of doing high-quality work more efficiently and thus improving productivity and its competitive position, it makes a difference whether 5 percent or 85 percent of the machine tools are under some stage of automated machine control and what proportion of the plant workload is allocated directly to those machines.

The question of implementation is no longer one of simple diffusion but one of *penetration* across the full range of distinct metalworking operations and of *saturation* of the total workload of the plant. Put simply, to what extent does an adopting metalworking plant now produce differently? The ultimate effect of NC/CNC is tied more directly to how widely it is used in a plant than to whether it has been adopted at all. Further discussion of saturation is offered in the following chapter. I now consider the extent to which NC/CNC has penetrated the inventory of machine tools in this segment of the industry.

Technology Penetration Trends. In any particular plant or shop, the combination of machine functions performed will be determined by a number of factors, including the supplier network of which it is a part, the contracts that traditionally sustain it, the market niche that it has carved out for itself, and the degree and nature of specialization that it imposes on itself as a result. Small plants are often known to be highly specialized and are sustained by dependence on relatively focused market niches. Consequently, the full range of machine functions is not usually found in any one plant.[12] Still, a predictable variety of machining and related metalworking operations are commonly found in plants in this sector. Among the more frequently encountered machine functions performed are metal turning, such as would be done on a lathe, and metal drilling, milling, and boring. Electrical discharge machining (EDM), in which an electrode does the cutting, punch pressing, and flame-plasma cutting, in which a torch is used to make the cut, are far less frequently encountered.

In the plants and shops in this study the more common machine operations performed were likewise found to include metal turning, drilling, milling, and boring. EDM operations, punch pressing, and flame or plasma cutting were reported far less frequently. Nonetheless, the degree to which these functions have come to be controlled by advanced machine control techniques as opposed to conventional machine control—which is another way of saying subject to the eye-hand coordination of a skilled machinist—should indicate the degree of rationalization in a particular plant. Our question becomes, Has a plant simply acquired a token NC/CNC capability, or has this advanced technology actually begun to leave its stamp on the bulk of the work that the plant performs?

Table 21 reports the number and percentage of plants with varying numbers of mechanized metalworking operations automated at least minimally. Generally, two broad categories of metalworking operations are present in these plants and shops. Machining centers, drilling, milling, and boring machines, and lathes are commonly found; punch presses, grinding machines, EDM, and inspection machines are quite rare. Furthermore, even where an operation is commonly performed in a plant, generally no more than three machines are controlled by advanced technologies.

A sizable proportion of plants and shops have only a very small number of machines performing a particular operation, and large proportions even of plants in which an operation is commonplace do not have a single operation automated. Consequently, these data serve as partial evidence of the low degree of penetration of machining operations in these plants by NC/CNC-CAM technology.[13]

To explore the issue of technology penetration inside plants in more detail, let us examine data on the proportions of specific machine operations that are automated at least minimally (table 22). The most important finding is that while CNC has generally overtaken NC machine control in plants that have stepped up from conventional to automated machine control, the penetration by NC/CNC-CAM of metalworking operations is still a mile wide and an inch deep. As recently as 1982, large shares of the machines actually performing in plants and shops in this segment of the metalworking sector were not yet materially influenced by the existence—and the commercial avail-

TABLE 21

Plants with Varying Numbers of NC/CNC Machine Tools, 1982

Type of NC/CNC Machine Tool	Number of Tools in Plant					
	1–3	4–6	7–9	10+	Not reported	Total
Machining centers	447 (38.1)	107 (9.1)	27 (2.3)	20 (1.7)	571 (48.7)	1,172 (99.9)
Drilling/boring/milling machines	581 (49.6)	88 (7.5)	24 (2.0)	29 (2.5)	450 (38.4)	1,172 (100.0)
Lathes	539 (46.0)	116 (9.9)	22 (1.9)	29 (2.5)	466 (39.8)	1,172 (100.1)
Punch presses	148 (12.6)	8 (0.7)	0 0	4 (0.3)	1,012 (86.3)	1,172 (99.9)
Grinding machines	47 (4.0)	8 (0.7)	5 (0.4)	4 (0.3)	1,108 (94.5)	1,172 (99.9)
Regular/wire EDM	70 (6.0)	11 (0.9)	1 (0.1)	0 0	1,090 (93.0)	1,172 (100.0)
Inspection machines	120 (10.2)	5 (0.4)	0 0	3 (0.3)	1,044 (89.1)	1,172 (100.0)

NOTE: Figures in parentheses are percentages. Percentages may not add to 100.0 because of rounding.

ability—of more sophisticated machine control technologies. For example, only 6.0 percent of all the plants that perform metal-turning operations had more than 50 percent of their lathes and related machinery under numerical control, and 18.3 percent had more than 50 percent governed by CNC. Half of all plants (49.8 percent) had more than half their metal-turning machines controlled conventionally. The same general pattern holds for drilling, milling, and boring machines. The most significant pattern in the data for regular or wire EDM, punch pressing, or flame or plasma cutting is that these operations are extremely rare among plants in this segment of the sector.

In general, the rate at which CNC has eclipsed NC in these plants has exceeded the rate at which any form of automated machine control has overtaken and displaced conventional control. As the degree of penetration increases from less than 10.0 percent to more than 50.0 percent of existing machines, progressively higher proportions of the machines are under the control of CNC than of NC. Yet this must

TABLE 22
TECHNOLOGY PENETRATION: MACHINES IN PLANT UNDER ALTERNATIVE TYPES OF OPERATION CONTROL, 1982
(percent)

Type of Operation	Type of Control	None	Less Than 10.0%	10.0–25.0%	25.1–50.0%	More Than 50.0%	Not Reported	Total
Metal turning	NC	307 (26.2)	42 (3.6)	128 (10.9)	88 (7.5)	70 (6.0)	537 (45.8)	1,172 (100.0)
	CNC	222 (18.9)	24 (2.0)	155 (13.2)	177 (15.1)	214 (18.3)	380 (32.4)	1,172 (99.9)
	CONV	31 (2.6)	55 (4.7)	179 (15.3)	191 (16.3)	584 (49.8)	132 (11.3)	1,172 (100.0)
Metal drilling, milling, or boring	NC	218 (18.6)	56 (4.8)	218 (18.6)	129 (11.0)	115 (9.8)	436 (37.2)	1,172 (100.0)
	CNC	182 (15.5)	40 (3.4)	206 (17.6)	214 (18.3)	219 (18.7)	311 (26.5)	1,172 (100.0)
	CONV	23 (2.0)	45 (3.8)	214 (18.3)	240 (20.5)	515 (43.9)	135 (11.5)	1,172 (100.0)
Regular or wire EDM	NC	433 (36.9)	4 (0.3)	5 (0.4)	7 (0.6)	14 (1.2)	709 (60.5)	1,172 (99.9)
	CNC	432 (36.9)	5 (0.4)	8 (0.7)	12 (1.0)	31 (2.6)	684 (58.4)	1,172 (100.0)
	CONV	393 (33.5)	8 (0.7)	11 (0.9)	15 (1.3)	115 (9.8)	630 (53.8)	1,172 (100.0)
Punch pressing	NC	405 (34.6)	4 (0.3)	27 (2.3)	24 (2.0)	51 (4.4)	661 (56.4)	1,172 (100.0)
	CNC	419 (35.8)	1 (0.1)	17 (1.5)	19 (1.6)	44 (3.8)	672 (57.3)	1,172 (100.1)
	CONV	294 (25.1)	13 (1.1)	69 (5.9)	33 (2.8)	229 (19.5)	534 (45.6)	1,172 (100.0)
Flame or plasma cutting	NC	440 (37.5)	—	3 (0.3)	4 (0.3)	12 (1.0)	713 (60.8)	1,172 (99.9)
	CNC	425 (36.3)	2 (0.2)	3 (0.3)	4 (0.3)	27 (2.3)	711 (60.7)	1,172 (100.1)
	CONV	351 (29.9)	9 (0.8)	14 (1.2)	7 (0.6)	179 (15.3)	612 (52.2)	1,172 (100.0)
Other	NC	4 (0.3)	—	2 (0.2)	1 (0.1)	1 (0.1)	1,164 (99.3)	1,172 (100.0)
	CNC	4 (0.3)	—	1 (0.1)	1 (0.1)	2 (0.2)	1,164 (99.3)	1,172 (100.0)
	CONV	4 (0.3)	—	—	1 (0.1)	3 (0.3)	1,164 (99.3)	1,172 (100.0)

NOTE: CONV = conventional. Figures in parentheses are percentages. Percentages may not add to 100.0 because of rounding.

be seen against a background defined by two important factors. First, relatively high proportions of plants do not perform certain types of machine operations. Second, where an operation is performed, it is highly likely that it will still be controlled conventionally—that is, by a skilled operator-machinist—rather than by either NC or CNC technology.

Temporal Patterns of Technology Adoption. What has been the patterning of the penetration of selected metalworking operations through plants, aside from the nominal adoption of advanced technology by plants, across time? Table 23 reports the patterns of adoption. We have already seen (table 20) that the diffusion into these metalworking plants and shops began slowly in the early 1960s and peaked most recently in the late 1970s. This same pattern is evident in table 23, although the inroads made by NC in metal turning and drilling, milling, and boring operations in the early 1970s are notable. The late 1970s saw the surge of CNC across these operations. Here we see the process of technology and to a lesser extent product life cycles and their succession at work most clearly. Nevertheless, for operations such as EDM, punch pressing, and flame or plasma cutting, virtually no significant inroads have yet been made, and the little that has occurred was confined to the period following 1975.

Factors Influencing the Adoption of Automated Machine Control Technology

The decision to adopt advanced manufacturing technology is doubtless governed by many factors. Although the combination of economic circumstances defining time periods may be ever changing, it is instructive to search for the enduring factors that prompt decisions to adopt and upgrade technology across time. Although NC technology has been in existence for over three decades, it did not filter down into the broad base of the metalworking sector until relatively recently. Even then, the extent of the diffusion has been less than spectacular. The succession of NC and then CNC systems into plants and their penetration through application to a range of operations and the capital equipment that performs them has been confined principally to the past decade. From the perspective of the larger metalworking sector, these patterns are far from signifying a major rationalization of the greatest proportion of the plants and shops that make up the sector. From the perspective of trying to understand the diffusion and penetration that have taken place, new questions present themselves.

What factors have influenced plants to make decisions to adopt new technology? Respondents in these plants were asked to rank-

94

TABLE 23

TEMPORAL PATTERNS OF NC/CNC DIFFUSION AMONG TYPES OF METALWORKING OPERATIONS, BEFORE 1950 TO 1982

Date of Initial NC/CNC Adoption		Metal Turning		Metal Drilling, Milling, and Boring		Regular or Wire EDM		Punch Pressing		Flame or Plasma Cutting	
		No.	CP[a]	No.	CP[a]	No.	CP[a]	No.	CP[a]	No.	CP[a]
Before 1950	NC	1	0.5	—	—	—	—	—	—	—	—
	CNC	1	0.2	—	—	—	—	—	—	—	—
1950-54	NC	1	0.5	1	0.2	1	3.1	—	—	—	—
	CNC	1	0.5	—	—	—	—	—	—	—	—
1955-59	NC	—	0.5	2	0.5	—	3.1	1	0.9	—	—
	CNC	—	0.5	—	—	—	—	1	1.2	—	—
1960-64	NC	5	1.8	39	6.9	1	6.3	7	7.4	1	4.8
	CNC	—	0.5	2	0.3	1	1.6	—	1.2	—	—
1965-69	NC	36	10.5	170	34.8	3	15.6	11	17.6	—	4.8
	CNC	3	0.5	—	0.3	—	1.6	2	3.6	—	—
1970-74	NC	146	47.8	176	63.7	5	31.2	30	45.4	4	23.8
	CNC	15	3.5	29	4.6	3	6.5	5	9.5	2	5.1

(Table continues)

TABLE 23 (continued)

Date of Initial NC/CNC Adoption		Metal Turning		Metal Drilling, Milling, and Boring		Regular or Wire EDM		Punch Pressing		Flame or Plasma Cutting	
		No.	CP[a]	No.	CP[a]	No.	CP[a]	No.	CP[a]	No.	CP[a]
1975–79	NC	177	92.7	185	94.1	14	75.0	42	84.3	6	52.4
	CNC	320	59.3	391	62.8	28	51.6	36	52.4	16	46.2
1980–82	NC	29	100.0	36	100.0	8	100.0	17	100.0	10	100.0
	CNC	233	100.0	250	100.0	30	100.0	40	100.0	21	100.0
Not reported	NC	777	(66.3)	563	(48.0)	1,140	(97.3)	1,064	(90.8)	1,151	(98.2)
	CNC	599	(51.1)	500	(42.7)	1,110	(94.7)	1,088	(92.8)	1,133	(96.7)
Total	NC	1,172	100.0	1,172	100.0	1,172	100.0	1,172	100.0	1,172	100.0
	CNC	1,172	100.0	1,172	100.0	1,172	100.0	1,172	100.0	1,172	100.0

a. CP = cumulative percentage. This figure across time periods denotes the cumulative proportion (that is, extent of diffusion) of all NC/CNC machine controls for a specific operation.

order seven factors commonly associated with such decisions. The overlapping and varying specificity of the factors was deliberate so as to permit different aspects of the same basic factor to be considered.[14] Of special importance are factors that contribute directly in different ways to desired improvements in productivity as well as those that imply the option of proceeding with a skilled labor *compensation* strategy where workers with sufficient skills (for example, experienced machinists) arc not available or with a labor *substitution* strategy where the appropriate labor is available at a wage rate judged to be prohibitive. It is important to recall that these data were gathered in late 1982 and early 1983 after nearly two years of official recession and an even longer period of curtailed activity for many plants in many industries. As I noted in chapter 3 and explore in greater detail in chapter 6, the urge to compensate for chronic shortages of skilled labor was probably considerably blunted by late 1982 since the inability to retain employees during the back-to-back recessions of the early 1980s, more than any shortage of labor, probably dominated the concerns of plant managers.

Table 24 shows how respondents rank-ordered the seven factors. Decisions to automate machine tool operations have been driven primarily by the more proximate desire to increase plant productivity. More than seven in ten (71.1 percent) of the respondents ranked this factor either first or second in importance.[15] The factor ranked first by the second highest percentage (20.1 percent) was the more distant objective of improving a plant's competitive position in its industry.

This overriding concern for increased productivity that can translate into improved competitiveness is particularly understandable given the specific competitive features of metalworking industries and how they have developed for more than a century and a half, especially since World War II. The importance of localized and regional markets is enhanced by the relatively high costs of transporting metal products. These plants and shops have a tendency to cluster near major industry customers, especially in the fabrication of end products; and medium-sized and small plants are thus especially sensitive and vulnerable to local and regional competition.

The desire to improve quality assurance was ranked first or second by the second highest percentage (37.0 percent) of the plants. To a considerable extent this factor may be considered instrumental to the larger effort of boosting productivity. The greater precision and reduced tolerances demanded by sophisticated metal fabrication production processes have increasingly shifted importance to quality assurance.

Despite the origins of NC technology in the early 1950s as the outcome of an early version of military-university cooperative R&D,

TABLE 24
RANKINGS OF FACTORS PROMPTING NC/CNC ADOPTION

Reason for NC/CNC Adoption		1st	2nd	3rd	4th	5th	6th or Lower	Unranked Factor	Not Reported	Total
Increase productivity	No.	518	315	127	38	9	2	96	67	1,172
	%	44.2	26.9	10.8	3.2	0.8	0.2	8.2	5.7	100.0
Improve quality assurance	No.	133	301	281	118	52	9	81	197	1,172
	%	11.3	25.7	24.0	10.1	4.4	0.8	6.9	16.8	100.0
Compensate for shortage of skilled workers	No.	74	122	147	133	121	94	54	427	1,172
	%	6.3	10.4	12.5	11.3	10.3	8.0	4.6	36.4	99.8
Improve competitive position within industry	No.	236	169	235	118	54	22	85	253	1,172
	%	20.1	14.4	20.1	10.1	4.6	1.9	7.3	21.6	100.1
Acquire state-of-the-art technology	No.	51	35	75	86	136	162	32	595	1,172
	%	4.4	3.0	6.4	7.3	11.6	13.8	2.7	50.8	100.0
Required by DOD/NASA contract	No.	3	1	10	6	7	169	1	975	1,172
	%	0.3	0.1	0.9	0.5	0.6	14.4	0.1	83.2	100.1
Control labor costs	No.	52	97	160	143	125	93	52	450	1,172
	%	4.4	8.3	13.7	12.2	10.7	7.9	4.4	38.4	100.0

NOTE: Percentages may not add to 100.0 because of rounding.

requirements related to military or defense contracts do not figure prominently in prompting the adoption of advanced technology. Although the technological sophistication of production processes is often formally stipulated in contracts let by the Department of Defense and the National Aeronautics and Space Administration for highly sophisticated parts, this reason for adoption is ranked very low, perhaps at least partly because few of these plants or shops either bid on or receive such contracts.

Finally, the desire to compensate for a shortage of skilled workers or to engage in a strategy of labor substitution to put downward pressure on or otherwise adjust to total labor costs does not figure prominently in the ranking of motives for adopting automated manufacturing technologies. One plant in six (16.7 percent) ranked the desire to compensate for a shortage of skilled workers either first or second; one in eight (12.7 percent) ranked the desire to control labor costs either first or second. Although these two factors rank relatively low, the plants for which they are especially salient appear to view them as indirect strategies for retaining a competitive position in an increasingly hostile economic environment.

These data must be interpreted in the context of the period during which they were gathered. The diffusion of NC and later of CNC through the metalworking sector—and their penetration through plant operations—was largely confined to the past decade. For the skeptic to suggest that plant managers might mask strategies calculated to contain or reduce labor costs by labor substitution through automation by stated desires of improving productivity and increasing competitiveness is understandable. Since, however, by early 1983 employment loss was widespread and the importation of finished products—like the imports of relatively inexpensive and high quality machine tools themselves—was taking a devastating toll on suppliers of parts to domestic producers, these findings can probably be taken at face value. The data offer every indication that this surge in automation was under way despite the cyclical downturn of the late 1970s and the long and deep recession of the early 1980s. These findings are at least consistent with the rationale that the increased automation of metalworking production held the promise of slowing the employment contraction already taking place and thereby saving jobs by improving competitiveness rather than reducing labor costs by deliberately eliminating jobs.

Technology Upgrading: CNC Programming Expansion Plans

Having examined the shift from conventional to some form of automated machine control, let us consider the strategy open to plants

99

that had already adopted at least minimal NC or CNC capabilities for upgrading technology by further computerizing their existing NC operations. Plans for expanding CNC programming capability can be a sensitive indicator of the differing ways in which plant managers read the economic environment and take steps to reposition their plants to survive extended recessions and intensified competition from foreign producers.

When asked about plans to expand CNC programming capabilities, sizable proportions of the plants in this segment of the industry reported that they planned not only to expand such capabilities but to do so relatively soon. Overall, 76.3 percent of the plants reported plans to expand their programming capabilities before the end of 1984. Although 54.1 percent of the plants with such plans had no specific target date for implementing them, 41.6 percent of the plants planned to do so during the first six months of 1983 and another 32.6 percent in the last six months of that year. These responses suggest that CNC programming expansion plans may have been motivated by active and present considerations along with considerable pent-up demand, again probably the result of uncertainty tied to past recessions.

Regional Patterns. The expansion plans reported above are generally replicated in all nine geographic regions (table 25). In all regions except the East South Central, more than 70 percent of all plants reported plans to expand their CNC programming capabilities. As in the patterns of original NC/CNC adoption, the significance of regional location pales in comparison with other factors that influence decisions to adopt or expand technology. Specifically, there is little evidence that the older industrial regions in which metalworking plants and shops have long been concentrated were laggard, recalcitrant, or otherwise incapable of taking the necessary steps to ensure their adjustment and survival. That lower proportions of plants in the East South Central, West South Central, Mountain, and Pacific regions had expansion plans may be because plants in those regions are among the newest in the nation and already had more expanded CNC capabilities than older plants in other regions.

Corporate Structure. Corporate structure was found to vary among regions. Might it not also exert an influence on a nonlocational aspect of production—the decision to upgrade already adopted automation technology? There is considerable policy concern that plants that are part of multiplant companies, especially branch plants that serve as satellite production centers, may resort to a strategy of subtle disin-

100

TABLE 25

PLANT PLANS TO EXPAND CNC PROGRAMMING CAPABILITIES,
BY REGION, 1982

Region	Yes		No		Total	
	No.	Percent	No.	Percent	No.	Percent
New England	86	82.7	18	17.3	104	100.0
Middle Atlantic	165	78.2	46	21.8	211	100.0
East North Central	298	78.0	84	22.0	382	100.0
West North Central	84	80.0	21	20.0	105	100.0
South Atlantic	54	80.6	13	19.4	67	100.0
East South Central	16	64.0	9	36.0	25	100.0
West South Central	48	70.6	20	29.4	68	100.0
Mountain	27	73.0	10	27.0	37	100.0
Pacific	103	74.6	35	25.4	138	100.0
Total	881	77.5	256	22.5	1,137	100.0

vestment to adjust to increasing competitive pressures in particular locations. This disinvestment can take the form of indefinite delays in the introduction of state-of-the-art production technology or the upgrading of already acquired technology. Such delays could indicate a longer-term corporate strategy of gradually downgrading a plant by shifting away from traditional product lines, by surrendering competitive advantages through neglect, or by phasing out an operation in a location that is vulnerable to mounting cost pressures for whatever reason.

Concern for what has been viewed as the increasing velocity of capital mobility in the service of deliberate strategies to disinvest from operations in plants located, for instance, in non-right-to-work states, in plants in which production workers are unionized, or in plants located in otherwise high-wage environments is often expressed by focusing on what is happening among branch plants and how the results can indicate larger corporate intentions. Whether a plant is independent or part of a multiple-site operation does not appear to influence plans to upgrade or expand CNC programming capabilities (table 26). Some 78.7 percent of the single-site plants planned to expand their programming capabilities, and 74.2 percent of plants tied to larger corporate structures likewise planned to do so.

Plant Age. In a similar manner, a plant's age—an important proxy for

TABLE 26

Plans to Expand CNC Programming Capabilities, by Plant Type, 1982

Plans to Expand	Single-Site Plant		Component of Multiplant Company		Total	
	No.	Percent	No.	Percent	No.	Percent
Yes	630	78.7	250	74.2	880	77.3
No	171	21.3	87	25.8	258	22.7
Total	801	70.4	337	29.6	1,138	100.0

NOTE: x^2 = 2.70; degrees of freedom = 1; not significant at .05 level.

the vintage of the production capital inside—might also be considered influential in the decision to upgrade manufacturing technology. Relatively high and comparable proportions of plants in all cohorts reported plans to expand their programming capabilities (table 27). Again, plant age per se appears to have little influence on this kind of technology-based industrial adjustment.

Union Status. Finally, does the union status of a plant that already has minimal NC or CNC capabilities influence plans to expand those capabilities? Unionization has no apparent influence on programming expansion plans (table 28).

Access Factors Governing the Adoption of Technology

It is reasonable to suggest that the attractiveness of upgrading NC/CNC-CAM capacities already present in plants is influenced by both market and related contextual factors and particular plant characteristics. As we have seen illustrated by the lag between innovation and diffusion, whether new technology exists is far less important than whether it is accessible. Accessibility implies more than simple availability; it also implies some expectation that the fit between the capabilities of automated manufacturing systems and the workload of a specific plant is such that some minimal standard of economic feasibility will be met. Will the value of expected benefits at least equal the necessary expenditures of time and money associated with the adoption of technology?

TABLE 27

PLANS TO EXPAND NC/CNC MACHINE CONTROL, BY AGE OF PLANT, 1982

Plant "Birth" Cohort	Age of Plant (years)	Yes		No		Total	
		No.	Percent	No.	Percent	No.	Percent
1800–1849	133+	1	100.0	—	—	1	100.0
1850–1874	108–132	3	100.0	—	—	3	100.0
1875–1899	83–107	16	100.0	—	—	16	100.0
1900–1909	73–82	11	61.1	7	38.9	18	100.0
1910–1919	63–72	15	75.0	5	25.0	20	100.0
1920–1929	53–62	33	78.6	9	21.4	42	100.0
1930–1939	43–52	36	83.7	7	16.3	43	100.0
1940–1949	33–42	93	80.9	22	19.1	115	100.0
1950–1959	23–32	159	76.4	49	23.6	208	100.0
1960–1969	13–22	253	78.6	69	21.4	322	100.0
1970–1979	3–12	209	72.3	80	27.7	289	100.0
1980–1982	<3	16	76.2	5	23.8	21	100.0
Total		845	77.0	253	23.0	1,098	100.0

Two considerations that quickly translate into costs and therefore function as barriers to upgrading technology are the prospects of lengthy training periods and a high purchase price. To the extent that the expenses associated with either or preferably both of these factors can be minimized, these two potential barriers will have been reduced.

TABLE 28

PLANS TO EXPAND NC/CNC MACHINE CONTROL, BY UNION STATUS OF PRODUCTION WORKERS, 1982

Plans to Expand	Union		Nonunion		Total	
	No.	Percent	No.	Percent	No.	Percent
Yes	184	75.1	676	77.7	860	77.1
No	61	24.9	194	22.3	255	22.9
Total	245	22.0	870	78.0	1,115	100.0

NOTE: $x^2 = 0.73$; degrees of freedom = 1; not significant at .05 level.

TABLE 29

Expected Length of Training Period

Hours	Number	Percent
1–5	19	1.6
6–10	79	6.7
11–20	92	7.8
21–40	393	33.5
41–80	173	14.8
81–120	108	9.2
121–160	20	1.7
161–200	31	2.6
200+	63	5.4
Not reported	194	16.6
Total	1,172	99.9

NOTE: Percentages do not add to 100.0 because of rounding.

Expected Length of Training Period. The expected length of time that it will take to train a machine operator or supervisor to become proficient in upgraded CNC programming technology can be a useful indicator of how economic feasibility is viewed by plant managers. To obtain some measure of the time commitment involved, plant managers were asked about the length of the training period anticipated to introduce and successfully to incorporate upgraded technology into plant operations. Nearly half the respondents (48.3 percent) expected such proficiency to take from more than one-half to two full weeks of training (table 29). For small and medium-size plants especially, this can be a menacingly large and inconvenient investment of time and labor resources and therefore a major barrier to upgrading. Programming systems that are more user friendly and can offer training in a shorter period, then, would seem to have a greater chance of being broadly adopted.

Expected Price. The expected price of such a system can present an even more tangible barrier to the adoption of upgraded technology by plants and shops. One in three plant managers (33.6 percent) anticipated that a computer-assisted programming system would cost between $15,000 and $50,000 (table 30). Since by 1982 it had become possible to purchase such a system for as little as $10,000, many metalworking plants in this segment may have been overpessimistic about cost considerations. This may explain why upgrading plans were less definite or extensive than they might otherwise have been.

TABLE 30

Expected Price of Computer-assisted Programming System

Expected Price	Number	Percent
Less than $1,000	15	1.3
$1,000–4,999	59	5.0
$5,000–9,999	82	7.0
$10,000–14,999	114	9.7
$15,000–24,999	145	12.4
$25,000–49,999	248	21.2
$50,000–99,999	68	5.8
$100,000–249,999	34	2.9
$250,000+	23	2.0
Not reported	384	32.8
Total	1,172	100.1

NOTE: Percentages do not add to 100.0 because of rounding.

In summary, I have traced in considerable detail the patterns of both actual and planned technology upgrading among plants and shops in the sample. Clearly, although there has indeed been a surge of adoptions of new NC/CNC technology in recent years, the implementation of that technology—as measured by penetration into actual machine operations—has lagged considerably behind its diffusion. The application of automated manufacturing technology has been quite narrow and limited. Since the major reasons reported for the diffusion have been related to the desire to increase productivity directly rather than to do so indirectly by using automation to control labor costs, the next chapter explores more fully the consequences of the adoption of new technology inside metalworking plants and shops.

Notes

1. See J. M. Utterback, "The Dynamics of Product and Process Innovation in Industry," in C. Hill and J. Utterback, eds., *Technological Innovation for a Dynamic Economy* (New York: Pergamon, 1979), pp. 40–65.

2. "The 13th American Machinist Inventory of Metalworking Equipment, 1983," *American Machinist* (November 1983); and Office of Technology Assessment, *Computerized Manufacturing Automation: Employment, Education, and the Workplace* (Washington, D.C., 1984), pp. 114, 279.

3. E. Mansfield, "The Diffusion of Eight Major Industrial Innovations," in N. E. Terleckyj, *The State of Science and Research: Some New Indicators* (Boulder, Colo.: Westview Press, 1977).

4. Mansfield has illustrated this well:

> In some industries, the improved electronics technology may not offer across-the-board economic advantages over the existing technology. For example, numerically controlled machine tools became available on a commercial basis in the mid-1950s. But the present state of the art offers economic advantages only for intermediate-size production lots, and not for large or small ones. Numerically controlled machine tools therefore took more than 15 years to capture 25% of machine tool sales, whereas in 15 years semiconductors achieved no less than 85% of the combined semiconductor/vacuum tube market.

Ibid.; quoted in N. Rosenberg and W. E. Steinmuller, "The Economic Implications of the VLSI Revolution," in N. Rosenberg, ed., *Inside the Black Box: Technology and Economics* (Cambridge: Cambridge University Press, 1982), pp. 178ff. See also N. Rosenberg, "Technological Changes in the Machine Tool Industry," *Journal of Economic History*, vol. 23 (December 1963), pp. 414–43.

5. "Computers in Manufacturing Use, Planned Use, and Buying Influence," *Production* (August–September 1982) (mimeo.).

6. The tendency for larger plants to adopt advanced production technologies at higher rates than smaller plants was also reported by J. Rees, R. Briggs, and R. Oakey, "The Adoption of New Technology in the American Machinery Industry," *Regional Studies*, vol. 18, no. 6 (1984), p. 493.

7. A relatively long lag between technology innovation and diffusion has been noted for such key innovations as digital and mini computers, basic oxygen furnaces in steel production, electricity generated by nuclear power, telephones served by electronic switching, and ocean-borne containerized liner cargo service. See Bureau of Labor Statistics, *Productivity Chartbook* (Washington, D.C., 1981), pp. 60–61. For an illustration of the relatively rapid diffusion of such innovations as automatic transmission, power steering, air conditioning, and disc brakes through the automobile industry, see discussion of W. J. Abernathy, *The Productivity Dilemma* (Baltimore: Johns Hopkins University Press, 1978), cited also in *The Competitive Status of the U.S. Auto Industry* (Washington, D.C.: National Academy Press, 1982), chap. 3, p. 48. For a discussion of the dangers inherent in confusing the innovation with the diffusion stages, see J. Alexander Smith, "Time Horizons and Inflation: A Premature Death for 'Postindustrial Society'?" *Southeastern Political Review* (Fall 1981), pp. 56–89.

8. Office of Technology Assessment, *Computerized Manufacturing Automation*, p. 280.

9. See N. Rosenberg, *Perspectives on Technology* (Cambridge: Cambridge University Press, 1976); and C. Freeman, J. Clark, and L. Soete, *Unemployment and Technical Innovation* (Westport, Conn.: Greenwood Press, 1982).

10. See L. G. Tornatzky et al., *The Process of Technological Innovation: Reviewing the Literature* (Washington, D.C.: National Science Foundation, 1983), pp. 83ff., chap. 8.

11. "13th American Machinist Inventory," based on data reported in table 8.

106

12. National Academy of Engineering, *The Competitive Status of the U.S. Machine Tool Industry* (Washington, D.C.: National Academy Press, 1983), p. 53.

13. The study of computerized manufacturing automation by the Office of Technology Assessment (*Computerized Manufacturing Automation*, p. 59) noted that the extent of penetration of NC/CNC machine tool control may be understated somewhat since older tools not currently in use are commonly counted in the machine tool inventory.

14. The list of factors was designed to overlap considerably with that used by Rees et al., "Adoption of New Technology."

15. Inevitably, the social desirability of potential answers can pose a problem to the extent that plant managers attempt to assign priority to reasons that they judge will meet with wide approval and conceal more base motives such as covert plans to eliminate costly labor. Such a response by survey respondents is exceedingly difficult to isolate and identify from data sets of any kind.

6

Automation, Employment, and Workers' Adjustment

In chapter 4 I examined a variety of industrial adjustments made by U.S. metalworking industries to shifting features of the larger economic environment during the past century and more. The gradual upgrading of physical capital through a succession of new plant cohorts, the retention of its concentrated regional location, and the shift of growth toward nonunionized labor environments illustrated the diversity of this adjustment process. These adjustments were interpreted as forms of mobility, whereby restructuring within the industry led certain features of the sector to grow more rapidly than others, with the result that the larger industry complex was continuously being repositioned in its economic environment. These same adjustments have shaped the metalworking sector both as a target and as a setting for yet another form of industrial adjustment—the automation of basic production—which was the focus of chapter 5.

The diffusion of advanced manufacturing technology through an industry is, however, only one aspect of the upgrading of technology. Although chapter 5 viewed the adoption of new production technology as a consequence of broader changes in the economy, it may be viewed as a cause of still others. The effects of automated machine control on the metalworking sector and the implications of this kind of adjustment for productivity, competitiveness, employment, and the world of work itself are the focus of this chapter.

Worker and Machine: Conflict and Coexistence

The key distinguishing feature of industrial economies has long been the mechanization of production. Machines—and the technologies they embody—have become increasingly important ingredients in combinations of factor inputs. The evolution of machines toward greater technical sophistication inevitably led to adjustments in the worker-machine interface. In work settings of all kinds, machines and mechanized production processes have become the principal organizing influences for work. The gradual blurring of the distinction between machines as extensions of workers and machines as replacements for them has brought about a major tension of industrial society.

Although ever more sophisticated technologies and their diffusion throughout entire industries are centuries old, the specter of massive job displacement has generated flurries of concern only intermittently. For the most part, increasingly automated production has enjoyed widespread support, largely because of the belief that rising living standards are a major byproduct. "Since the turn of the century, the introduction of new technology has generated more capital, added to the number of people employed and increased the average income of each employee."[1] Moreover, tracing the links between the elimination or redefinition of traditional occupational specialties in otherwise expanding industrial economies has only occasionally attracted interest. Besides, the celebrated misgivings of the Luddites and the analyses of mechanized production by others notwithstanding, historical evidence shows that automation has generally been induced by incentives to minimize production time rather than to eliminate labor.[2] The labor-saving potential of factory automation may be far less important than the ability to use increasingly expensive and exotic production materials more efficiently.[3]

The past quarter-century has been bracketed by two flare-ups of anxiety about the effects of automation on the workplace. Not surprisingly, both were triggered by distinct life stages of the computer. In the late 1950s and early 1960s, the computer as an invention and the rampant—if premature—speculation about its potential applications set off the first of these modern controversies. The dust settled quickly, in large part because the nation was beginning a long period of economic expansion during which any unsettling effects of automation were more than compensated for by continued employment and income growth. The second of these flare-ups began in the early 1980s and continues today. It, too, is largely a response to the computer, although now the points of debate concern actual, not anticipated, effects of the diffusion of the computer microprocessor in its myriad forms throughout society, especially the workplace.[4] The productivity slowdown and the stagnant employment growth in manufacturing—including absolute job loss in several industries during the early 1980s—lent high visibility to the prospect of machines replacing men and women on factory floors around the nation, even though in reality the slow pace of automation has probably rendered the issue moot for most industries.

What some observers see as a second industrial revolution is

109

thought to be poised to sweep across the factory floor with the same ferocity as the first swept through a primarily agrarian and extractive economy beginning two centuries ago. Not only are the full range of production activities vulnerable to increased pressures for greater productive efficiency, but the world of work itself is vulnerable as well. It has been estimated that automation could bring about a 20–25 percent reduction in the factory work force over the next ten years.[5] Yet in the metalworking sector this general prophecy may need to be more carefully qualified. Given the chronic shortage of skilled machinists for conventional machine control operations, even a future surge of automation is not expected to generate as much job displacement in machining as in fabrication and assembly.[6]

Even so, not until the recent recession and the forecast of a permanent loss of jobs did the full weight of concern about job displacement by manufacturing automation come to be felt. For some observers the falloff of investment in capital equipment after the 1960s appeared to belie any accelerated trend toward greater automation.[7] Yet the visibility of factory closings and the estimates that private disinvestment by American business caused the loss of 30–40 million jobs in the 1970s alone[8] have revived anxiety over the role of automation in industrial change. Widespread contraction of employment throughout an industry that is viewed as a symptom of a restructuring regional, national, or international economy may well induce a plant—whether it is expanding or contracting its own employment—to take steps to guard against future signs of diminished competitiveness.[9] It is in this way that unorchestrated industrial adjustments can lead to profound social consequences.

It can be argued that factory automation has not yet exacted a toll on existing metalworking employment because so little diffusion of new and advanced process technologies has taken place. This view suggests that the effects will be fully felt only when factory automation begins to be more widely diffused. In the present global economy, however, new technology may not even have the opportunity to eliminate jobs. The jobs may already be gone because of the inability of much of U.S. metalworking to compete in international and domestic markets against strong and efficient foreign producers. Ironically, then, future factory automation may actually save jobs. Illustrative of this reworked causal sequence between technological change and employment effects is a recent government report on robotics, which notes that the adoption of such productivity-enhancing technologies as robots and other forms of automation has generally occurred *after* the industry has experienced declines in employment.[10]

Nevertheless, the more conventional suspicion is that automation can be used by plant managers to trim costs through labor substitution and thus purposely to eliminate jobs, to change their contents or composition on the shop floor, or otherwise to reduce dependency on highly paid skilled labor and thereby increase managerial control.[11] Because this possibility undoubtedly exists, the adoption of advanced technologies is commonly associated with job destruction. A counter-argument is that on the whole and in the long run—two qualifications that understandably try the patience of many—the savings in unit production costs that can be realized by more technologically advanced production arrangements and associated organizational changes may just as easily translate into lower prices, expanded markets, and therefore eventual job creation.[12]

Although the causal sequence between technology adoption and employment change is probably not fixed for all times, places, or industrial circumstances, it is reasonable to view either secular or cyclical employment contraction as a sufficient incentive for a company to consider upgrading its manufacturing facilities. In seeking to explore the employment consequences of such a strategy, I turn first to what has been happening to metalworking employment in general.

Patterns of Metalworking Employment Change, 1980–1982

The industries that make up the U.S. metalworking sector sustained significant employment losses in the early 1980s, although the severity of those losses was far from uniform across individual industries. Far more net employment loss was caused by metalworking establishments' going out of business than by their inability to hire more workers than they laid off. In the absence of more detail, this suggests that business failures, not just the accompanying employment loss, were particularly troublesome during the recent recession. Employment losses were inversely related to firm size, the largest firms sustaining the heaviest losses.

This pattern of severe employment contraction is also found in the sample data collected for this study. Roughly three of four plants in the sample experienced net employment declines during 1980–1982. What was the structure of this widespread employment loss? The net effects of employment expansion and contraction and establishment formation and failure among metalworking shops during the early 1980s redistributed plants among categories of employment size. The general pattern was of downward mobility to smaller size categories. In 1980 some 11.3 percent of the plants had one to ten production employees, and 7.6 percent had total employment in that range (table 31). By 1982, 17.5 percent of the plants had production employ-

TABLE 31
PLANT EMPLOYMENT, 1980 AND 1982

| | 1982 | | | | 1980 | | | |
| | Production | | Total | | Production | | Total | |
Number of Employees	No.	Percent	No.	Percent	No.	Percent	No.	Percent
None	4	0.3	—	—	3	0.3	1	0.1
1–10	205	17.5	137	11.7	132	11.3	89	7.6
11–20	209	17.8	168	14.3	189	16.1	142	12.1
21–50	309	26.4	288	24.6	296	25.3	292	24.9
51–100	228	19.5	231	19.7	232	19.8	208	17.7
101–250	104	8.9	202	17.2	150	12.8	230	19.6
250+	35	3.0	85	7.3	48	4.1	105	9.0
No response	78	6.7	61	5.2	122	10.4	105	9.0
Total	1,172	100.1	1,172	100.0	1,172	100.1	1,172	100.0

NOTE: Percentages may not add to 100.0 because of rounding.

ment and 11.7 percent had total employment in that range.

The swelling of the ranks of small plants at the expense of larger plants is evident for both production and total employment. The primary reason for the shift was that employment losses caused many plants to fall into the next smaller size category. Since the sample consists of a fixed group of plants doing business both in 1980 and 1982, the net effects of plant formation and failure are not registered in these data. As noted in chapter 3, however, the expansion of the ranks of plants with small work forces was attributable not only to employment losses in existing plants but also to the larger number of formations than of failures among small plants in all metalworking industries in the early 1980s. Since the two dates closely bracket the recessions of the early 1980s, it is possible that both short- and long-term economic adjustments are confounded in accounting for this employment contraction. Since the time period is so short, however, the effect of larger structural changes is effectively controlled.

The Illusion of Plant Age. Is there any evidence that the employment contraction was influenced by the age distribution of metalworking plants? That is, is there any reason to suggest that employment changes differed between younger and older plants? The data indicate that older plants were more likely than younger plants to lose employment (table 32). Of the plants of 1920s vintage, for example, 95.2 per-

TABLE 32

EMPLOYMENT CHANGE BY AGE OF PLANT, 1980–1982

Year Plant Began	Age of Plant (years)	Increase		Decrease		Total	
		No.	Percent	No.	Percent	No.	Percent
1800–1849	133+	—	—	1	100.0	1	100.0
1850–1874	108–132	—	—	3	100.0	3	100.0
1875–1899	83–107	1	6.3	15	93.8	16	100.1
1900–1909	73–82	2	12.5	14	87.5	16	100.0
1910–1919	63–72	2	11.1	16	88.9	18	100.0
1920–1929	53–62	2	4.8	40	95.2	42	100.0
1930–1939	43–52	7	17.9	32	82.1	39	100.0
1940–1949	33–42	18	16.5	91	83.5	109	100.0
1950–1959	23–32	38	20.0	152	80.0	190	100.0
1960–1969	13–22	94	30.8	211	69.2	305	100.0
1970–1979	3–12	81	31.3	178	68.7	259	100.0
1980–1982	<3	8	47.1	9	52.9	17	100.0
Total		253	24.9	762	75.1	1,015	100.0

cent lost employment during 1980–1982. The proportion of the next six plant cohorts that experienced net employment losses during 1980–1982 dropped consistently, to only 52.9 percent for plants in the most recent cohort. On the surface, at least, this suggests that plants that are more likely to embody older capital stock may also offer less efficient—and more costly—production settings. Their diminished competitiveness would probably be reflected in labor costs that stubbornly resist economizing at stable employment levels. Recession, then, required the trimming of employment in these plants.

In chapter 5 we saw that even older plants showed a capacity to upgrade their production technology. Does this pattern of diffusion of NC/CNC-CAM through these plants in the late 1970s run counter to an explanation of employment loss by plant age? Perhaps not. Even though advanced production technologies such as NC/CNC-CAM made inroads into those plants, relatively few metalworking operations in the plants had been automated by the end of 1982. Ultimately, productivity effects flow from the extent of technological implementation within plants, not the extent of diffusion to plants. Moreover, typically only small proportions of total plant workloads were affected by the new technologies; simple diffusion appears to have outdistanced the penetration or saturation processes in the automation of metalworking operations. In the end, the apparent influence of plant age probably masks the fact that younger plants, which experienced relatively less employment loss, faced markets in regions in which employment declines were less prevalent.

Regional Location. The employment contraction exhibited a noticeable regional patterning (table 33). Net declines in production employment were most prevalent among plants in the East North Central region, in which the metalworking sector is the most heavily concentrated. More than eight plants in ten had to release production workers during the two-year period. In addition to diminished demand, the cutting back by the heavy concentrations of basic manufacturing industries in this region on orders for parts from small subcontractors tightened markets for the metal products suppliers. Smaller and fewer orders spread over a fixed number of plants set the stage for significant layoffs.

The proportions of plants whose employment declined in the same period exceeded 70.0 percent in all regions but two—the South Atlantic and the East South Central. This suggests that the location of firms in the southern rim insulated them from the full effect of the contraction. That the Southeast is dominated by states with right-to-work provisions and traditionally strong antiunion sentiment, which together may offer a more favorable labor cost environment for business, cannot be easily dismissed. For this sample of plants, however, it is more likely that the regional growth associated with in-migration and rising personal income—especially in manufacturing and construction—raised demand levels and thereby reduced competitive pressures in the local markets of these metalworking plants. Whether this might also indicate a conscious channeling of business by multilocational firms to branch plants in these relatively attractive business climates cannot be determined from these data. Although this possibility is fully consistent with the data, the relatively small extent of interregional branching among metalworking firms in this sample argues against the significance of this factor. Interregional outsourcing (that is, securing parts from producers in cheaper markets) by larger firms to which the sample firms are tied through backward linkages remains a possibility, however. I turn to a more detailed consideration of these possibilities below.

The Influence of Labor Environment

At a time of heightened concern over rapid capital mobility and corporate flight in search of attractive business climates, the influences of unionization and labor environments favorable to it on employment change attract considerable attention. If the conventional wisdom—or

114

TABLE 33
Employment Change by Region, 1980–1982

Region	Increase No.	Increase Percent	Decrease No.	Decrease Percent	Total No.	Total Percent
New England	29	29.3	70	70.7	99	100.0
Middle Atlantic	53	27.9	137	72.1	190	100.0
East North Central	62	18.5	273	81.5	335	100.0
West North Central	27	28.4	68	71.6	95	100.0
South Atlantic	25	39.1	39	60.9	64	100.0
East South Central	8	33.3	16	66.7	24	100.0
West South Central	15	24.2	47	75.8	62	100.0
Mountain	9	27.3	24	72.7	33	100.0
Pacific	29	22.8	98	77.2	127	100.0
Total	257	25.0	772	75.0	1,029	100.0

the preceding analysis—is any guide, we should not be surprised that unionized plants in this sample bore the brunt of the employment loss generated during 1980–1982. Between 1980 and 1982, 85.6 percent of the unionized plants and 71.7 percent of the nonunion plants experienced employment declines (table 34). Since the vast majority of unionized plants are located in non-right-to-work states of the Northeast and Midwest, it is somewhat surprising to find that the prevalence of employment loss was not significantly greater in non-right-to-work states than in right-to-work states (table 35). While 76.0 percent of the plants in states without right-to-work provisions experienced a loss of production jobs, 70.1 percent of plants in states with right-to-work provisions also did so. This suggests that the distinction between a state legislative environment that either promotes or inhibits unionization and the actual unionization of production employees is particularly useful here. When attempting to understand the structure of employment change, the influence of the labor setting that is more proximate to the plants and shops actually experiencing the change is

TABLE 34
EMPLOYMENT CHANGE BY UNION STATUS OF PRODUCTION WORKERS, 1980–1982

Employment Change	Union		Nonunion		Total	
	No.	Percent	No.	Percent	No.	Percent
Increase	34	14.4	223	28.3	257	25.1
Decrease	202	85.6	565	71.7	767	74.9
Total	236	23.0	788	77.0	1,024	100.0

NOTE: χ^2 = 18.68; degrees of freedom = 1; significant at .001 level.

TABLE 35
EMPLOYMENT CHANGE BY RIGHT-TO-WORK STATUS OF STATES, 1980–1982

Employment Change	Right-to-Work State		Non-Right-to-Work State		Total	
	No.	Percent	No.	Percent	No.	Percent
Increase	52	29.9	207	24.0	259	25.0
Decrease	122	70.1	656	76.0	778	75.0
Total	174	16.8	863	83.2	1,037	100.0

NOTE: χ^2 = 2.69; degrees of freedom = 1; not significant at .05 level.

more relevant and more easily traced.

The Degree of Employment Change

Although the direction of employment change (gain or loss) provides a general indication of the effects of industry characteristics on employment, nothing has yet been said about the degree or extent of such employment effects and the patterns that describe them. From an analytical perspective, it is as useful as it is wise to differentiate between degrees of severity of employment change, especially losses.

Table 36 reports the influence of union status and corporate structure on the direction and degree of employment change between 1980 and 1982. Clearly, the direction was influenced by whether a plant was unionized. Even after adjusting for the fact that nonunion plants are in the clear majority in this sample, we find that among plants in which employment expanded, nonunion plants captured the bulk of the gains. There was a higher probability that a job gained would go to a nonunion plant and that a job lost would be suffered by a union plant. Nonunion plants captured more employment gains of 10 percent or more, while unionized plants were more likely to have captured net employment gains of less than 10 percent. Nonunion plants were also more likely than union plants to have experienced employment stability during 1980–1982. Unionized plants were more likely than nonunion plants to have been affected by loss of employment, and the tendency was replicated in the majority of levels of severity.

The data offer no evidence of a relationship between the direction of employment change and whether a plant is a single-site establishment or a component of a multiplant corporate structure. Regardless of the direction or the degree of employment change, the change experienced by single-site plants was not substantially different from that experienced by multiestablishment firms.

A Closer Look at Employment Change

In the following analyses, a series of regression models were tested to explore patterns of change in production employment in the sample of metalworking plants and to verify the earlier analyses.[13] Three independent variables were introduced to account for changes in production employment between 1980 and 1982: (1) UNION, whether a plant's production workers are unionized; (2) PLANT TYPE, whether a plant is independent or part of a more complex corporate structure; and (3) PLANT AGE. Three models were specified and examined to see how well they fitted the data.

In the first model, the focus was on whether the unionization of a plant's production workers accounts for employment change (YP) during 1980–1982.[14] The results indicate that while employment loss was very common throughout the metalworking sector, losses were especially widespread in unionized plants. The intercept (−8.81) of the regression line indicates an average employment loss of 8.8 percent among nonunionized plants during 1980–1982. If we can assume that few laid-off employees were later recalled during those two years, approximately 8.8 percent of the employees in nonunionized plants lost their jobs during that time. The slope of the regression line measures the difference between the average employment change in unionized plants and the average in nonunionized plants. That the

117

TABLE 36
Employment Change by Union Status of Production Workers and Plant Type, 1980–1982

Employment Change	Union		Nonunion		Total		Single-Site Plant		Component of Multiplant Company		Total	
	No.	%	No.	%	No.	%	No.	%	No.	%	No.	%
Growth (%)												
Greater than 100.0	1	5.0	19	95.0	20	100.0	12	60.0	8	40.0	20	100.0
75.1–100.0	1	6.3	15	93.8	16	100.1	9	56.3	7	43.8	16	100.1
50.1–75.0	—		18	100.0	18	100.0	16	88.9	2	11.1	18	100.0
33.4–50.0	4	8.7	42	91.3	46	100.0	29	63.0	17	37.0	46	100.0
20.1–33.3	8	12.7	55	87.3	63	100.0	46	73.0	17	27.0	63	100.0
10.1–20.0	11	17.2	53	82.8	64	100.0	49	76.6	15	23.4	64	100.0

5.0–10.0	6	25.0	18	75.0	24	100.0	16	61.5	10	38.5	26	100.0
Less than 5.0	3	50.0	3	50.0	6	100.0	3	50.0	3	50.0	6	100.0
No change	23	18.3	103	81.7	126	100.0	96	76.2	30	23.8	126	100.0
Loss (%)												
Less than 5.0	1	20.0	4	80.0	5	100.0	4	57.1	3	42.9	7	100.0
5.0–10.0	14	33.3	28	66.7	42	100.0	30	73.2	11	26.8	41	100.0
10.1–20.0	38	28.6	95	71.4	133	100.0	85	63.9	48	36.1	133	100.0
20.1–33.3	38	21.8	136	78.2	174	100.0	133	76.4	41	23.6	174	100.0
33.4–50.0	50	31.4	109	68.6	159	100.0	111	67.7	53	32.3	164	100.0
50.1–75.0	35	33.7	69	66.3	104	100.0	68	64.8	37	35.2	105	100.0
75.1–100.0	3	12.5	21	87.5	24	100.0	18	75.0	6	25.0	24	100.0
Total	236	23.0	788	77.0	1,024	100.0	725	70.2	308	29.8	1,033	100.0

effect of the recent recession fell particularly heavily on unionized plants can then be clearly inferred. Above and beyond the average employment loss experienced by nonunionized plants, the loss attributable to union status was an additional 13.8 percent, for a total of 22.6 percent. That is, the average unionized plant lost more than one production worker in five. The figures in the parentheses are the *t*-values for the intercept and the slope. Since the values are highly significant, this model shows a strong relationship between employment change and unionization.[15]

It appears that union status was a particularly heavy burden for metalworking plants to bear during the recessionary 1980–1982 period. High wage-benefit packages and restrictive work rules and job classifications are the specific features of unionized labor settings generally assumed to be responsible for the relative inflexibility of unionized plants when faced with changing economic circumstances. To the extent that these are features of the unionized plants in this sample, they may have inhibited shifts to more productivity-enhancing technology. Their influence may also be reflected in the production and managerial environments in such a way as to exacerbate competitive disadvantages. In good times the extra burden of these arrangements is borne with far less trouble than when the economy slows down. During times of cyclical downturn and longer-term structural adjustments in the economy, unionized production arrangements may aggravate such problems and indirectly cause job losses.

In a second model, the variable *PLANT AGE* was added to the first model. The significance of both *t*-values indicates the existence of relationships between employment change and both union status and plant age and thus confirms the earlier findings.[16] Not only did employment loss afflict unionized plants severely, but it was greater in older than in younger plants.[17]

Once again, given the possibility of retrofitting older plants with advanced production arrangements, the age of a plant may not be directly related to employment change. But age may imperfectly reflect more subtle contextual factors, including the industrial traditions that define it and the local-regional markets to which it is anchored. Metalworking plants do not experience employment losses because the capital arrangements that house them are old, but older plants are often more vulnerable to a kind of institutional arthritis from which younger plants may be relatively insulated. The inability of a plant to adapt to new economic circumstances could manifest itself in a reluctance to consider the adoption of advanced manufacturing technologies to help it retain its competitive position even though the larger industry is constantly shifting in the direction of ever newer physical plant.

Older plants, concentrated as they often are in intensely competitive industrial subsectors and regional markets, should have all the incentives they need to update their plant and equipment and so retain or restore their ability to compete. Yet they may have lost their agility and may either underestimate the survival value of upgrading their capital equipment and less tangible managerial styles or not be prepared to do so in a timely fashion.

Finally, the third independent variable, *PLANT TYPE*, was introduced into the model.[18] This model is not an improvement over the leaner second model. No strong linear relationship appears to exist between the complexity of corporate structure and employment change. Single-site plants were neither more nor less likely than headquarters and branch plants to experience job gains or losses during the 1980–1982 period, thus confirming the earlier analysis.[19]

Responding to Employment Change

The plants and shops from which the data for this study are derived all share at least two characteristics. Each has 250 or fewer employees and each has already acquired at least some minimal NC or CNC machine control capability. In many cases that capability is relatively insignificant or tied to only a few machines. Nonetheless, even relatively small increments of upgraded technology can be associated with characteristic ways in which the new technology is accommodated or leaves its imprint on the internal workings of the production environment. This section seeks to identify and trace those effects.

The widespread employment contraction experienced by metalworking plants in the sample—particularly older and unionized ones—during 1980–1982 may be viewed as a consequence of key features of the industry and the economic context in which it functions. It may also serve as an incentive for actions to lessen an industry's vulnerability to employment contraction and the loss of competitiveness such widespread contraction reflects. Employment loss, then, may be viewed as both a cause and an effect. In this section I explore the possibility that the experience of severe employment contraction can prompt plants and shops to make plans to enhance and expand whatever automated manufacturing capacities they already have.

Shifting employment levels—dominated as they are by contraction—are related to plans to expand NC/CNC-CAM capabilities within metalworking plants (table 37). Fully 77.0 percent of the plants reported plans to expand their computer-assisted (that is, CNC) programming capabilities. Although the pattern is replicated among both plants that gained and plants that lost employment during 1980–1982, plants that gained employment were more likely to have plans to

121

TABLE 37

PRODUCTION EMPLOYMENT CHANGE BY PLANS TO EXPAND
NC/CNC MACHINE CONTROL, 1980–1982

Plans to Expand	Increase		Decrease		Total	
	No.	Percent	No.	Percent	No.	Percent
Yes	211	83.1	577	74.9	788	77.0
No	43	16.9	193	25.1	236	23.0
Total	254	24.8	770	75.2	1,024	100.0

NOTE: $\chi^2 = 7.13$; degrees of freedom = 1; significant at .01 level.

expand CNC programming than plants that had experienced employment decline. In large part, this finding is unexpected; it was anticipated that plants in which employment had declined would have a greater incentive to improve their productivity and competitiveness. Instead, those plants relatively unscathed by employment decline were more likely to report plans to expand technology. How might this be explained? One possibility is that in increasingly competitive local or regional markets, the coming shakeout, which was already under way before the recessions of the early 1980s, had a momentum against which plants already weakened and losing ground expected to make little headway even if they could afford to enhance their automated manufacturing equipment. By contrast, more competitive plants may well have decided that any future gains must depend on increased technological sophistication throughout their operations and management.

Accommodating Manufacturing Automation inside Plants

This section focuses on ways in which the small and often tentative steps toward automating production processes have been integrated in metalworking plants. How program preparation is handled, who on the shop floor is generally assigned the task, and the extent to which such tasks have become significant features of established workloads and production routines (that is, "saturation") are questions that help identify the range of effects of new automated technologies on older industrial work settings. Are programming tasks added to the responsibilities of existing plant personnel, or are they assigned to new and specially trained workers? Do plants attempt to handle the new tasks themselves, or do they establish subcontracting links to outside specialists to handle them?[20] These and related questions are explored here.

Our conclusions may not only prove instructive in themselves but also suggest what kinds of new technologies these plants and shops would find particularly attractive and useful and the conditions under which widespread adoption would be most likely to proceed rapidly. Knowledge of the initial effects of technology in these often low-technology production settings, no matter how minimal the extent of penetration, may offer clues to the future effects of technology on the work environment and workers in metalworking industries.

Method of Program Preparation. Workpiece program preparation entails differing commitments of time and combinations of labor skills, depending on the sophistication of the automated machine tool control system. The relatively low sophistication of the NC/CNC-CAM capabilities in this sample of plants and shops is easily seen. Of the various ways in which NC/CNC workpiece programs are usually prepared, the most common is manual: 41.6 percent of the plants prepare their workpiece programs by hand (table 38). Yet 56.9 percent of the plants use computer assistance in preparing all or part of the necessary workpiece programs.

Programming as an Occupational Task. One of the most commonly anticipated influences of advanced manufacturing technology on plant operations is its potential for redefining—to the point of rendering obsolete and redundant—the job descriptions and skill criteria of older occupational hierarchies.[21] Let us take a closer look at effects of this kind. What is the occupational niche of the persons generally responsible for preparing NC or CNC workpiece programs?

Traditionally, the skilled craftsmen in these settings were those who could work manually or with tools guided by their hands with high accuracy and consistency. Increasingly, the adoption of automated tool guidance systems has necessitated a redefinition of the skills considered essential in a so-called skilled worker. The desired accuracy and the consequent improved productivity become an outcome less of trained eye-hand coordination directly than of the ability to instruct and guide machine tools indirectly through programmable machine tool control systems. For some observers this constitutes a form of deskilling; yet it may just as easily be seen as a shift of emphasis from historically valued skills toward a new constellation of skills more consistent with the capabilities of the new technology. It is not so much that, on a continuum between man and machine, the balance shifts toward the machine as that the worker must add new

TABLE 38
ALTERNATIVES FOR NC/CNC WORKPIECE PROGRAM PREPARATION

Alternative	Number	Percent
Manually	488	41.6
With computer assistance	268	22.9
Combination of both	398	34.0
No response	18	1.5
Total	1,172	100.0

skills to his repertoire to retain mastery over the machine. In this process traditional occupations are recast and redefined, and new ones are created. Of course, how the redefined occupations are allocated to the existing labor force is at the heart of much skepticism over such automation.

Given the relatively modest sizes of the plants in this study and the related constraints on specialization of tasks on the shop floor, it would be reasonable to expect the responsibility for workpiece program preparation to fall to someone other than a full-time programmer. Yet the person responsible for CNC workpiece program preparation is most often a full-time programmer (table 39). In 21.1 percent of the plants the programming responsibilities are handled by the machine operators, in 19.3 percent by shop supervisors, and in 16.6 percent by part-time programmers. Outside specialists are resorted to very infrequently (0.9 percent). As the new technology is adopted, it appears to become integrated into older production arrangements in such a way as to require either the broadening of the skills of the traditional machinist or new skilled workers altogether. What apparently does not happen is a buildup of dependence on external specialists.

Type of Computer Support in Use. It is possible that the upgrading of production technology can prompt plants to enter into subcontracting relations with external agents to meet their new computer support needs.[22] Given the often minimal penetration of CNC capacities among plants in this sample, we might expect this tendency to be minimal. Some 39.2 percent of the plants use in-house computer

124

TABLE 39

PERSON MOST OFTEN RESPONSIBLE FOR NC/CNC WORKPIECE PROGRAM PREPARATION

Program Preparer	Number	Percent
Machine operator	247	21.1
Full-time NC/CNC programmer	389	33.2
Outside contractor	10	0.9
Shop supervisor	226	19.3
Part-time NC/CNC programmer	194	16.6
Other	87	7.4
No response	19	1.6
Total	1,172	100.1

NOTE: Percentages do not add to 100.0 because of rounding.

facilities to support their computer-assisted machine control operations (table 40). No doubt reflecting the widespread availability of microcomputers in many forms, only 17.4 percent of the plants secure such support either from outside vendors or from a company mainframe computer located in another plant.

"Saturation" of the Programming Workload. It has been suggested that the shift of manufacturing to small volumes—Toffler's "mouse milk" runs—sets the stage for great growth in flexible automation.[23] Moreover, it has been reported that 75 percent of all machined parts are handled in batches of fifty or fewer.[24] As batch manufacturing eclipses mass manufacturing, the flexibility accompanying soft-wired programmable machine tool control should be especially attractive, but this may be truer of relatively large plants than of the smaller ones that are the focus of this study.

Given the frequent specialization of plants in this stratum of the sector as subcontractors to and suppliers for other industries and the small proportion of machine tools in most plants under automated control, the weekly programming load by any one plant is likely to be light. For the most part, the data support this conclusion. Nearly four

TABLE 40
Type of Computer Support in Use

Programming Source	Number	Percent
On-site mainframe	115	9.8
Micro/mini	345	29.4
Shared time on company computer located elsewhere	57	4.9
Time purchased from outside vendor	146	12.5
Other	27	2.3
No response	482	41.1
Total	1,172	100.0

plants in ten (39.1 percent) prepare only two to five workpiece programs per week, and 75.0 percent prepare ten or fewer (table 41).

This limited requirement for reprogramming may reflect an important feature of production inside these plants. The structure of work in them is dominated by moderate-length production runs as well as a narrow spectrum of products. That suggests that this segment of the metalworking sector, composed of thousands of small and specialized plants and shops, is characterized by a production environment at some variance with that of the much smaller stratum of larger firms. Typically, the range of machining operations in smaller plants is narrow, and the operations required are repetitive and predictable rather than ad hoc or variable. Furthermore, only one person is usually preparing workpiece programs at any one time (table 42). Thus the range and volume of workload operations may not be perceived as sufficiently burdensome to generate a significant demand for programmable production technologies to lighten the load. For these plants this would stand as a monumental barrier to the shift from conventional or even basic NC to programmable machine tool control (CNC).

Complexity of Programming Tasks. Not only do these small plants have generally light workpiece preparation workloads, but perceptions of the complexity or difficulty of manufacturing tasks may offer

TABLE 41
Average Weekly NC/CNC Workpiece Program Work Load

Programs Prepared in Average Week	Number	Percent
None	8	0.7
1	147	12.5
2–5	458	39.1
6–10	266	22.7
11–15	89	7.6
16–50	142	12.1
51–100	11	0.9
More than 100	4	0.3
Not reported	47	4.0
Total	1,172	99.9

NOTE: Percentages do not add to 100.0 because of rounding.

clues to why saturation has not proceeded any further than it has. The data are somewhat ambiguous, however. Most of the metalworking operations in these plants are not perceived as difficult to program

TABLE 42
Average Number of People Preparing NC/CNC Workpiece Programs at Any One Time

Number	Number	Percent
None	8	0.7
0.1–1.0	830	70.8
1.1–3.0	288	24.6
3.1+	25	2.1
No response	21	1.8
Total	1,172	100.0

TABLE 43
NC/CNC Manufactured Parts
Difficult to Program

Percentage	Number	Percent
None	104	8.9
1–10	471	40.2
11–20	195	16.6
21–50	254	21.7
51–75	50	4.3
76–100	39	3.3
No response	59	5.0
Total	1,172	100.0

(table 43). While only 7.6 percent of the plant managers reported that more than half of their machining operations were difficult to program, another 8.9 percent noted that none of their workload tasks presented programming difficulties. These findings may be interpreted as indicating that plant managers see little need to incur the cost of upgrading technology since such small proportions of their workloads would benefit. The data also are consistent, however, with the opposite interpretation—since nine of ten plant managers have encountered at least some difficulty in parts programming with existing NC/CNC-CAM equipment, they have a strong disincentive to invest further in the technology. From this latter perspective, the feasibility of further and more extensive technology upgrading will depend on the development of more user-friendly control systems that require less formal training and eliminate common programming bottlenecks.

Conclusion

What does the evidence add up to? First, it does not suggest that a little advanced technology is necessarily responsible for generating the conditions required to support the demand for more advanced technology. Herein may lie a major reason for the unimpressive diffusion of advanced manufacturing technology through this segment of the metalworking sector, insignificant penetration across existing plant operations, and limited saturation of plant workloads in recent years. The light workpiece program loads, the relatively uncomplicated programming tasks in some settings, the possibility in others that existing numerical control capacities are not sufficiently trouble free, and the light manpower demands that accompany the level of numerically controlled machine tools already operating in these plants seem to diminish the incentive for plant managers to establish or upgrade their computer-assisted programming capacities.

Yet a large proportion of these plants are planning to do just that and to do it soon. Where is the prod suddenly coming from? It is fair to suggest that the need to overcome drawbacks of the existing installed technology may not figure as prominently as other, more subtle factors in the decision to upgrade technology. Automated production technology may make its next surge into these plants largely on the strength of the belief that they have no other choice and that plants that do not automate face a harsher future than plants that do.

Notes

1. Business-Higher Education Forum, "The New Manufacturing: America's Race to Automate" (Washington, D.C., 1984), p. 22. This quotation summarizes the following references: H. A. Simon, *The New Science of Management Decision* (Englewood Cliffs, N.J.: Prentice-Hall, 1977), chap. 5; Organization for Economic Cooperation and Development, *Technical Change and Economic Policy* (Paris: OECD, 1980); and W. Leontief, "The Distribution of Work and Income," *Scientific American*, vol. 247, no. 3 (September 1982), pp. 188–204.

2. B. Bluestone and B. Harrison, *The Deindustrialization of America: Plant Closings, Community Abandonment, and the Dismantling of Basic Industry* (New York: Basic Books, 1982); W. Rybczynski, *Taming the Tiger: The Struggle to Control Technology* (New York: Viking Press, 1983); and D. Kennedy, C. Craypo, and M. Lehman, eds., *Labor and Technology: Union Response to Changing Environments* (State College: Department of Labor Studies, Pennsylvania State University, 1982). See E. S. Ferguson, "History and Historiography," in O. Mayr and R. C. Post, eds., *Yankee Enterprise: The Rise of the American System of Manufactures* (Washington, D.C.: Smithsonian Institution Press, 1981), pp. 1ff.

3. For a discussion of the potentially misplaced emphasis on the labor savings of manufacturing automation, see Office of Technology Assessment, *Computerized Manufacturing Automation: Employment, Education, and the Workplace* (Washington, D.C., 1984), pp. 26, 33, 80.

4. Ibid. Computer-based automation may well give us an opportunity to rethink the traditionally rigid relationships between automation and employment change. As Corrigan has noted: "It can be argued that computer-based technology is the least deterministic, the most flexible technology to affect the workplace since the beginning of the industrial revolution." R. Corrigan, "Beyond the Bottom Line," *National Journal*, June 6, 1983, p. 1290.

5. See "The Factory of the Future," *Newsweek*, September 6, 1982.

6. G. Bylinski, "The Race to the Automated Factory," *Fortune*, February 21, 1983.

7. Michael Wachter, quoted in H. F. Myers, "Economists Say Slump Has Hastened Some Trends but Spawned Very Few," *Wall Street Journal*, May 25, 1983.

8. Bluestone and Harrison, *Deindustrialization of America*, p. 9.

9. Lawrence notes the acceleration of structural change across regions during the 1950–1980 period. R. Z. Lawrence, *Can America Compete?* (Washington, D.C.: Brookings Institution, 1984). For the regional implications of factory automation within geographically concentrated industry sectors, see R. U. Ayres and S. M. Miller, *Robotics: Applications and Social Implications* (Cambridge, Mass.: Ballinger Publishing, 1983); and H. A. Hunt and T. L. Hunt, *Human Resource Implications of Robotics* (Kalamazoo, Mich.: Upjohn Institute for Employment Research, 1983).

10. Joint Economic Committee, *Robotics and the Economy*, Staff study (Washington, D.C., 1982).

11. Dun's *Business Month*, February 1984, p. 17; and D. F. Noble, *The Forces of Production: A Social History of Industrial Automation* (New York: Knopf, 1984).

12. Joint Economic Committee, *Robotics*.

13. This section is derived from analyses prepared by Akbar Torbat, a graduate research assistant in the Political Economy Program, University of Texas at Dallas.

14. The resulting fit of this first model to the data yielded the following equation:

$$YP = -8.81 - 13.81 \; UNION \qquad R^2 = .02$$
$$(-5.97)\,(-4.50)$$

15. Since *UNION* is a dichotomous variable, the inherent restriction of its capacity to vary limits its ability to account for variation in the dependent variable, *YP*, which is measured as a continuous variable. Moreover, despite the significance of *UNION* in the equation for predicting change in production employment, we must be cautious in assuming a direct relation between the two. It is very likely that correlates of plant unionization more closely linked to stagnant productivity and lack of competitiveness may be generating a spurious relationship.

16. The resulting fit for the second model is as follows:

$$YP = -3.05 - 9.39 \; UNION - 0.26 \; PLANT \; AGE \qquad R^2 = .03$$
$$(-14.20)\,(-2.86) \qquad\quad (-3.65)$$

17. A related model, which tested for the influence of the combined effect of union status and plant age beyond their independent effects, yielded evidence of a nonsignificant relationship between employment change and the interaction of these two variables. This interaction term was therefore dropped from further analyses.

130

18. The resulting fit for this model is as follows:

$$YP = -3.76 - 10.4 \; UNION - 0.26 \; PLANT \; AGE + 3.17 \; PLANT \; TYPE$$
$$(-1.64) \quad (-3.04) \qquad\qquad (-3.58) \qquad\qquad\qquad (1.07)$$
$$R^2 = .03$$

19. Despite the significance of the relationships between two of the independent variables and employment change, the overall models are badly fitted to the data, as indicated by the low R^2 values. A likely reason for this is that both union status and plant type are dichotomous variables. The restricted range of their variation probably diminishes their capacity to covary across the full range of the dependent variable. Plant age, a continuous variable, does not appear to enhance the value of R^2, perhaps because this variable is associated with a very large variance term. The average plant age in this sample is twenty-five years, and the associated standard deviation is twenty years. Despite the poorness of fit to the data, however, the F-tests (t-values) reveal statistically significant relations between union status and plant age and the dependent variable production employment change.

20. For a discussion of the potential of programmable automation to bring about a new division of labor on the shop floor, see Office of Technology Assessment, *Computerized Manufacturing Automation*, p. 191. See also B. H. McCrackin, "Why Are Business Services Growing So Rapidly?" *Economic Review* (Federal Reserve Bank of Atlanta) (August 1985), pp. 14–28, which shows that the rise of the producer services reflects far more than simply the transfer of business and professional service occupations and activities out of the goods-producing sector.

21. See Office of Technology Assessment, *Computerized Manufacturing Automation*, chaps. 4, 5; and Noble, *Forces of Production*.

22. For a discussion of company decisions to contract out for services to support programmable automation, see Office of Technology Assessment, *Computerized Manufacturing Automation*, p. 151.

23. A. Toffler, *The Third Wave* (New York: Bantam Books, 1980).

24. Bylinsky, "Race to the Automated Factory."

7
Major Findings and Policy Implications

For more than a decade industrial transformation in the United States and abroad has been the focus of a great deal of attention. Much has been written concerning whether and how the United States can retain its half-century-long industrial dominance in a newly forming international economy. Predictably, varied interpretations have emerged to make sense of it all. One of the early and more compelling of these used the imagery of new "sunrise" industries slowly eclipsing older "sunset" industries. Gradually, however, the "cars-to-computers" simplicity of such an interpretation has been discovered to be as incomplete and misleading as it is beguiling. Regrettably, this perspective has obscured the fact that inefficient production processes and arrangements, not whole industries, are the chess pieces of a national economy and therefore the ultimate targets of change. Inevitably, the resulting policy debates reflected the power of this perspective to shape our understanding of what was happening. By the late 1970s policy questions concerned how and whether to try to identify industrial winners and losers, what might constitute the appropriate means and ends of an industrial policy, and what was the wisdom of pursuing public policies dedicated to guiding a nation's or even a region's industrial evolution.

As the limitations of this early view became evident, a second perspective arose as a partial corrective to the first. Attention slowly shifted to the importance of changes that might lead to the renewal, rather than simply the replacement, of older industries. This perspective assigned great importance once again to new technologies, but now to underscore their importance both to production arrangements and to final outputs as potential guarantors of industrial competitiveness rather than as the wellsprings of entirely new industries per se. The key questions became those that could help us identify the likely sources of technological innovation and the barriers to its adoption and diffusion that might prevent or delay necessary industrial adjustments. By and large, the problems faced by young and old firms in new and traditional industries were traced to both macroeconomic and microeconomic phenomena, including difficulties in capital formation and insufficient incentives to invest in new plant and equipment. As the 1980s began, the course of political change intersected with that of global industrial-economic change. The policy responses included the Economic Recovery Tax Act of 1981, which established tax credits and accelerated depreciation allowances to stimulate new capital investment and reduced the capital gains tax to 20 percent to encourage the accumulation of the billions in venture and expansion capital deemed necessary to sustain both a cyclical recovery and structural renewal throughout many of our older basic industries.

Generally this is where things stand today. With the recovery entering its fourth year, it is somewhat easier to isolate the effects of recession from those of restructuring—that is, cyclical from structural changes. Despite the unevenness of the current expansion, it is apparent that contemporary views of industrial adjustment continue to assign great importance to the role of new technologies. The legacy of our shifting understandings of and responses to industrial change will undoubtedly influence where we go from here. Therefore, the motive for this study has been to examine in greater detail this second perspective on industrial change, in which technological innovation for renewing, rather than emerging, industries plays such an outsized role. To this end I have examined industrial adjustments as they have unfolded among thousands of medium-sized and smaller firms in the more than 200 industries that constitute the U.S. metalworking sector. What have been the major findings, and what are their policy implications?

The Limits of Technological Transfer and Capital Investment

In this study, technology diffusion and implementation have been moved to center stage. Technology upgrading as a calculated intrafirm strategy of industrial adjustment includes not only the substitution of automated machine tool control systems—numerical control (NC-CAM)—for conventional arrangements (that is, manual control by a skilled machine operator) but also the upgrading to computer programmable control systems (CNC) by plants and shops that had already adopted earlier forms of automated machine control. The diffusion of automated forms of machine control through the metalworking sector has been slow and uneven across industries and machine functions, despite a recent rapid acceleration in adoption rates. More-over, the limited—if accelerating—diffusion has masked the extremely small effect on what actually goes on inside firms. What inferences can be drawn from these patterns?

Technological innovation is better viewed as a loosely coupled sequence of processes than as a discrete event. This sequence extends from the invention stage through the innovation stage, where the focus is on commercialization and patterns of application. The adoption-diffusion stage then directs attention to the patterns by which new—and often not-so-new—technologies are introduced into new settings. The diffusion of innovation *to* a firm sets the stage for its movement *through* a firm. The implementation stage includes a focus on the extent to which the full range of production activities inside a plant come to be affected by innovations (penetration) and the eventual share of the total firm output that is affected (saturation). Finally, the reception (accommodation/acceptance) of technological innovation by workers and managers must be considered part of the process of technology transfer.

Getting new technology into a plant is less important than getting it positioned throughout a plant and then rethinking the overall production process substantially, if not completely, so that increased shares of a plant's total output can be affected. Diffusion without implementation will have precious little relevance to industrial adjustments designed to restore competitiveness to firms in an older, beleaguered industry.

In light of this extended sequence, federal and state tax provisions designed to encourage increased investment in new capital plant and equipment have at best limited relevance for industrial renewal. In this analysis bringing greater conceptual clarity to discussions of technology transfer has not been an end in itself. Rather, doing so permits us to see that each stage may well be differently responsive to the policy tools that have recently received so much attention. Consider, for example, the implications of these findings for the investment tax credit (ITC). In the past quarter-century, the ITC has been far from a reliably constant feature of our federal tax code, having been variously liberalized, suspended, repealed, and reinstated over the years. Nevertheless, the ITC can doubtless constitute under certain circumstances a major inducement for firms to install new production technologies. An indirect reflection of this effect is that domestic machine tool orders have clearly risen and fallen in step with the status of the ITC since 1960.[1] Even though the net effect may well be to reduce the cost of such an investment by as much as a fifth, however, the relationship between new investment and industrial renewal is complex and indirect at best. The ITC may help get new technology introduced into a plant without necessarily ensuring that it will be implemented broadly or intensively once inside. Moreover, there is no assurance that the requisite rethinking of management support systems will take place after adoption. The diffusion of new technology is only a necessary, not a sufficient, prerequisite for the more consequential implementation stage.

Today, as we debate once again the fate of the ITC and accelerated depreciation schedules—key features of leading tax reform proposals—it is apparent that firms in many industries wisely place far greater emphasis on the potential for new markets, rising demand, and the general condition of the economy. Many firms in older industries may view the reduction of the corporate income tax rate and the continuation of the R&D tax credit as more consequential for industrial renewal than provisions aimed at capital formation and investment.

Adoptions of new technology generally take place in response to a changing economic climate rather than to changing technological or political—or political-economic—ones. The timing of the surge in NC/CNC-CAM adoptions underscores this point. Data in chapter 5 indicate that during the 1960–1964 period, a full decade after basic numerical control technology was commercially available, a discernible trickle of diffusion of numerically controlled machine tools was finally evident among medium-sized and smaller metalworking firms. The later availability of CNC made advanced forms of automated machine tool control systems increasingly accessible to and feasible for more of these plants and shops, but the diffusion did not peak until the 1975–1979 period, the last full period covered by the data.

Although such a time lag is commonly found in industrial diffusion, the timing of this lag is important for another reason. The bulk of the diffusion and even the later surge preceded the back-to-back recessions of the early 1980s and thereby occurred before any explicit debates over the wisdom of industrial policies, the alteration of tax incentives to spur new investment in plant and equipment, or even the widespread public recognition of the inroads being made by strong foreign competitors in domestic and foreign metalworking markets. Therefore, the timing of this surge can caution us against assuming that older industrial sectors are necessarily composed of economic actors that are somehow incapable of making the adjustments necessary to reposition themselves in a changing economic landscape. There is likewise little support for the notion that the mere availability of new technology is sufficient to trigger its widespread adoption.

New Tests of Corporate Rationality

Under certain circumstances, failure to adopt new technologies may be just as rational as adoption. Increasingly, it is assumed that a major test of corporate rationality should be an eagerness—or at least a willingness—to upgrade production technologies to the highest level of sophistication available. Indeed, in light of the exalted and even revered role assigned to technological innovation in recent years, how do we account for the fact that conventional machine tool control systems still dominate the production activities of U.S. metalworking plants and shops? A variety of explanations have been advanced.

Like many manufacturers, metalworking firms are extremely sensitive to interest rate and business cycle fluctuations. One result may well be an inability to plan for longer periods. It is also widely assumed that the managerial regimes of metalworking firms are relatively provincial and slow to appreciate and adopt changes. It is therefore tempting to explain the limited extent of technology upgrading by this presumed organizational or managerial stodginess. Similarly, to a certain extent what has been said of the domestic producers of machine tools can also be said of the metalworking firms that purchase those tools.[2] Examples abound of metalworking firms that exhibit a "small business mentality" even when facing expanding markets, are wedded to older, top-down authoritarian management styles and labor relations models, and eschew active involvement or interest in either in-house or external R&D programs. Yet at best these may be partial explanations; at worst they may be entirely misleading.

The limited degree of technology upgrading may be an entirely rational response to the specific characteristics that define production in medium-sized and smaller metalworking firms. Older production technologies and work arrangements will be retained if they are judged to be sufficient. The market for automated metalworking production in many small firms may be analogous to that for personal computers in the home. Justifying technology upgrading—especially computer-programmable machine tool control—for batch or longer runs of a narrow and specialized product line may be as difficult as justifying computerized checkbook balancing, address books, and recipe storage at home. A large gap remains between appropriate technology and optimal technology. And this distinction may even be greater for small-scale operations. For smaller firms the promises of new technology may be accompanied by proportionately higher risks. Although the adventuresome spirit among smaller firms has received much attention in recent years, the high failure rates among small businesses may well be substantially traceable to such unwarranted risk taking.

136

Ultimately, the criteria for rational corporate behavior among smaller firms may not differ from those governing larger corporate behavior so much as the margins for error do. Just as no apparent technological imperative propels a new technology into and through new industrial settings and applications, these metalworking firms have no inherent propensity to scale new technological heights simply "because they are there." Where that does happen, for many smaller firms, with their typically arhythmic cash flow, the impulse to commit resources to new and expensive process technologies may reflect a quest for a *deus ex machina* that can accelerate or even trigger their decline. In the long run corporate reluctance to automate may have survival value that has not yet been fully appreciated.

At the same time, the difference between a "hidden" corporate rationality and the absence of one can be very small. Smaller firms may well be no more—and may even be far less—capable than larger ones of anticipating changing market conditions, gathering and organizing economic intelligence, and acting on it in a timely and effective manner. The data presented in chapter 6 are consistent with this view. Consider these firms' responses to the downturns of the late 1970s and early 1980s. Severe, widespread, and protracted employment loss can serve as a powerful incentive to consider calculated strategies for regaining competitiveness in regional markets and beyond. Yet plants that reported the greatest interest in expanding their automated manufacturing technology were those that were relatively less affected by employment loss. On the one hand, this finding suggests that declining competitiveness may be a downward spiral and severe employment loss may be as enervating as it is demoralizing. Bad times reduce the range of options with which to stage a recovery and limit both the material and the managerial resources available for technology upgrading and its implementation. On the other hand, it may be testimony to the profound dependency of medium-sized and smaller metalworking firms on the fortunes of larger firms for which they often serve as parts suppliers. The rediscovery of small businesses in recent years, in which many commentators have ascribed to them a certain natural agility in negotiating treacherous market shifts and exploiting new market niches, may be an artifact of looking only at those that continue in existence. The dependency of individual small firms tied into complex intercorporate supply linkages may be such that they possess little capacity to extricate themselves from deteriorating economic circumstances. When business is bad for major producers, entire networks of small business satellites producing intermediate inputs go down with them.

As we look to the future, the circumstances accounting for the relatively low degree of diffusion and implementation may once again be in flux. Metalworking plants and shops reported plans to expand their CNC programming capabilities in the immediate future. Not only was this response widely reported from all regions of the nation, but whether a plant was tied to a larger corporate structure—as a branch plant would be—did not influence its reported plans to upgrade machine control technologies. Moreover, neither a plant's age nor whether its production workers were unionized influenced its upgrading plans. The uniformity of this response suggests that the stage may once again be set for a new surge of technology diffusion and implementation. The extent to which this actually occurs, however, is another matter and is an appropriate topic for future research.

The absorption of new process technologies alone cannot be expected to rescue individual firms as completely as it can be expected to assist in the adjustment of larger industries to new business conditions. The rationality of technology upgrading may be discovered as much in the replacement of business failures by business formations as in the calculations of individual firms making the difficult transition to new production arrangements. Consequently, the gravitation to new production arrangements tied to new process technologies holds greater promise for restoring or retaining the competitiveness of an industrial sector than for the survival of any particular firm. In the larger process of industrial renewal, the fortunes of an entire industry can be viewed as distinct from those of the firms that compose it. Industry health is entirely consistent with high rates of business failure provided that successive cohorts of new business formations are better able to meet the challenges of a changing business environment than those that they have replaced. A major portion of the technology upgrading in metalworking may well be accomplished by the coming on line of new firms with the latest technologies rather than by the massive retrofitting of older firms. To the extent that this is true, the adoption of new process technology within firms—and the policy tools created with that goal in mind—may not be as important to industrial renewal as other, aggregate-level adjustments accomplished through the replacement dynamics among firms. In other words, policy tools dedicated to preserving economic conditions supportive of high rates of economic growth and new business formation may be more consequential for overall industrial adjustment than well-intentioned "industrial policies" dedicated to easing the adjustments of individual firms.

Reappreciating Aggregate Industrial Adjustments

Conscious and calculated adjustments by firms, including the adoption of new technologies, may ultimately be of lesser importance to industrial renewal than relatively unorchestrated aggregate adjustments. Industrial change is a layered phenomenon. The full range of industrial adjustments includes those unfolding simultaneously within firms as well as within entire industries, sectors, and the larger economy. What happens at the level of individual firms can largely be viewed in isolation from what happens at more aggregate levels, if only because the rationality and calculated behavior so often judged to be the survival skills of modern business management play little if any direct role in more aggregate-level changes. Therefore, much industrial change is better viewed as the net result not only of what existing firms do but of the changing composition of an industry or sector.

Old industries are not necessarily composed of old firms housed in old production settings and employing outdated production technologies. Adjustment processes in an industry, a sector, and the larger economy depend on a ceaseless turnover of existing firms as illustrated by the data analyzed in chapter 4. The U.S. metalworking industry complex has been continuously and substantially restructured through plant cohort replacement and turnover throughout this century. The majority of medium-sized and small metalworking plants operating in 1980–1982 began operations after 1960. This process by itself provides a broad avenue for industrial renewal to the extent that the proportion of plants with new production technologies in place is greater among new plants than among plants that have gone out of business. Although even plants and shops that have existed for many decades—relatively rare in the metalworking sector—often retain a capacity to adjust adroitly to changing circumstances, that is not the point I am making here. It is important to appreciate that obsolescence is more a managerial and technological phenomenon than a calendrical one, but industrial renewal may proceed apace even in the absence of alert and adept management and the failure to retrofit existing plants with upgraded production technologies in a timely manner. Accordingly, the source and implications of the renewal process in a sector are quite separate matters.

A second indication of significant industrial adjustment at this level is to be found not in the upgrading of physical plant per se but in the transition to new labor environments within plants. Data in chapter 4 revealed that the steadily higher proportions of new plants that are nonunion have led to a long-term abandonment of unionized production arrangements. The declining influence of unions appears to be a result less of the decertification of unions in existing plants or the relocation of existing plants to avoid them than of new business formations that omit them. This shift may be viewed by some as an ominous trend, whereby the superior mobility of unfettered capital can gradually erode the institutional safeguards established by labor over the past half-century. But to the extent that the traditional safeguards for workers and labor relations functions have been transferred to new employee-management structures, legislatures, and courts, this adjustment may be viewed as not necessarily disadvantageous to labor. I turn to effects on labor in the following section.

Finally, the filtering into new physical plant and production settings and the filtering out of older labor relations arrangements were accomplished without abandoning older regional industrial concentrations. The fear that industrial adjustment would take place through runaway plants or extensive branching into regions offering lower labor costs and less union-oriented business climates seems not to have been substantiated. The aggregate shifts examined in chapter 4 were discovered to have taken place without reducing or displacing historical regional concentrations in the East North Central and Middle Atlantic regions. An important implication of this trend is that the specter of entire regions of the nation unable to revitalize themselves is largely unsupported. Indeed, the limited usefulness of the concept of "region" is amply illustrated by the ability of major portions of the U.S. metalworking sector to abandon older physical plant and labor environments without also abandoning older historical regions.

Technology Transfer and Industrial Policy Reconsidered

Inevitably, these observations encroach upon the policy debate surrounding the advisability of one or another form of explicit industrial policy by which an advanced economy might consciously attempt to guide its own industrial evolution. Most variants of industrial policy place great emphasis on encouraging rapid technological development within and between industries. It is worth the effort to rethink this approach. The technology innovation-diffusion-implementation sequence is of undeniable significance to industrial change and the effort to retain or regain competitiveness. In the end, however, permanent changes are being registered on industries throughout the U.S. metalworking sector as a result of their encounters with new global economic facts of life and a restructuring of domestic demand for their myriad products. There is little prospect that any conceivable "technological fix" or combination of fixes can stave off adjustments throughout this sprawling sector involving employment, sales, and market shares.

Far from being a universal antidote, the adoption of new automation technologies must be seen for what it is—an *economic* response to new circumstances. The inescapable implication is that technology

140

upgrading may make sense for some firms in some industries without making sense for other firms in other industries. Our assessment of successful industrial adjustment must allow for these differences and not impose the expectation that new production technologies are the sine qua non of industrial change. The choice of responses and their timing are perhaps best left to individual firms, if only because there is little evidence that any external "industrial policy" agent could guarantee superior results in guiding the larger process of industrial adjustment.[3] Along the way many individual firms may well misinterpret their own positions and options in this changing context without the larger industry's being placed in jeopardy. And, ultimately, it is the health of the industry and the larger national economy of which it is a part, rather than the longevity of specific firms, that commands the greater public interest.

Employment Effects and Workers' Adjustment

Technology concerns more than just hardware inputs and outputs of production operations; it includes also the functions that tools serve to improve organizational performance, and the interactions that tools have with their social setting.[4]

The employment effects of lost markets are likely always to be more debilitating and certain and less easily accommodated than the employment effects of automation and related forms of technology upgrading. The adoption of advanced production technology as both the consequence and the cause of employment and work-setting-related circumstances was the focus of chapter 6. Employment loss was widespread during 1980–1982, affecting three-fourths of the national sample of plants in this study. One consequence was that metalworking plants shifted into smaller employment categories. Metalworking employment loss was more often caused by firms going out of business than by layoffs exceeding new hires in plants and shops that stayed in business. Older plants were more likely to experience employment losses than newer ones, a finding that invites our examining far more carefully what correlates of "age" might account for it. Regional location also exercised an influence. Net declines in production employment were more common in the East North Central region than elsewhere. Nonetheless, except for the South Atlantic and East South Central regions, employment contraction affected more than 70 percent of the plants and shops in all regions during 1980–1982.

Unionized plants experienced significantly greater employment loss than nonunionized plants regardless of whether they were located in states that had passed right-to-work legislation. This seems to suggest that "business climate" is influenced less by a state's legal orientation toward open shops than by actual perceptions of the influence of organized labor inside plants. Finally, whether a plant was a single-site enterprise or part of a larger corporate structure did not influence the employment change it experienced during 1980–1982. Amid much concern for the presumed vulnerability of local (and regional) economies sustained by branch plants, it appears that such plants were not more vulnerable to employment losses during cyclical downturns. Branch plant status provided neither more nor less exposure to employment loss than single-site plant status. That the regional distributions of metalworking branch plants and single-site plants are not substantially different may help explain why employment losses by branch plants were not larger than those sustained by main production facilities or single-site plants, as is often thought to be the case in other manufacturing sectors.

Technology adoption and implementation are not the end points in an upgrading process. They may better be viewed as stages in a more circular process, in which they make adaptations by production workers and managers necessary. Employment effects can be viewed both as responses to the spread of new automation technologies and as reasons for it. New technologies present employees with a changed work environment as well as reordered production sequences. New skills, combinations of skills, and organizational adaptations are often necessary to accommodate new technologies. If resistance inside the plant is too great, implementation is threatened, and new technologies may not be permitted to have the desired effects. Inevitably, new technologies can alter more than simply familiar production sequences within plants and shops in an industry.

How are skill combinations and occupational hierarchies on the shop floor affected by the adoption of new programmable machine tool control systems? Ultimately, the specter of negative effects of automation on the workplace and workers has to be understood at this level. The empirical evidence offered in this study is neither extensive nor conclusive, but it does suggest that the dominant reasons for employment loss have long been a declining global market share caused by high-quality and low-cost foreign producers, a restructuring of domestic demand in metal product markets, and outmoded management and production arrangements rather than direct or widespread displacement of workers by automation. Moreover, new NC/CNC-CAM production technologies have been accommodated less by systematic "deskilling" than by reordering and even expanding the skills of existing workers. The task of workpiece program preparation most commonly falls to a full-time programmer and next most often to a machine operator. Furthermore, there is scant evidence that firms in these strata turn to external service support firms for either computer support through time sharing or workpiece programming assistance and thereby reduce on-site employment opportunities.

A partial explanation for the minimal influence on existing labor and production arrangements on the shop floor is the low degree of saturation that has taken place. Quite small proportions of the total workloads of these metalworking plants and shops are controlled by automated machine tool control systems. Moreover, as noted in chapter 5, anticipation of heavy initial investments in time and implementation costs and their disruptive effect on ongoing production schedules were found to influence the perceived feasibility of technology upgrading.

Another reason also suggests itself as being of considerable significance. Generally, the degree of specialization of plants in these strata is so great that there is only a limited requirement for reprogramming machine tool operations frequently, a capacity that may be more attractive to a producer of more diverse products. Production operations in most metalworking plants are commonly limited and repetitive rather than variable. Moreover, either they are relatively simple when they must be done at all, or they remain sufficiently difficult despite existing stand-alone NC-CAM capacities to discourage further investment in automated control systems. As a result, plants have often had little incentive to adopt programmable control systems. Plants where very small lots of customized or otherwise specialized runs are performed often have little incentive even to replace manual machine tool control arrangements. Thus the structures of workloads of metalworking plants that function as satellite subcontractors for major downstream producers are often such that the gains offered by automated machine tool control systems may be quite small.

Retargeting Policy Responses to Industrial Change

Technology upgrading understood as conscious retrofitting of older production arrangements in the U.S. metalworking sector inevitably takes place against the backdrop of larger and less visible industrial adjustments. Together these trends provide an important context in which to understand the extent to which any technology transfer has—and has not—occurred. In this study, these contextual factors have included industrial restructuring principally through new busi-

143

ness formations, failures, expansions, contractions, and rarely reloca-tions. The U.S. metalworking sector has been adjusting in dramatic and important ways all through this century quite independently of the evolution and diffusion of advanced manufacturing technologies.

What policy guidelines are available to those interested in whether, and if so how, the U.S. economy and the regional economies that compose it can adjust to new competitive circumstances? I may begin answering this question by suggesting that not only may indus-trial renewal not rest on new technology upgrading as much as com-monly thought, but it may also rest less on rational corporate calcula-tions and behavior by existing firms than is commonly thought. Indeed, much significant industrial adjustment takes place as new businesses replace older ones. This locates a significant portion of industrial renewal between firms rather than within them and nar-rows the range of potentially significant effects—positive or negative—associated with management acumen and public policies aimed at spurring capital investment and paving the way for technology trans-fer. Industrial adjustment is only partly and imperfectly a result of conscious, coherent, and coordinated actions. Relatively unorchestra-ted industry adjustments appear to provide more avenues for the kinds of technological and nontechnological adjustments that can lead to eventual industrial renewal.

Underscoring the potential for significant industrial adjustments through plant turnover need not be interpreted as stating a preference for a laissez-faire approach to thinking about and responding to in-dustrial change. The benefits of attempting to resurrect the myth of free markets in a modern economy are seriously outweighed by the costs of doing so. Rather, if we acknowledge that the replacement of older plants by newer ones may lead to substantial industrial adjust-ment, including rising technological sophistication of production ar-rangements, we should rethink the tools and targets we use to influ-ence industrial change. Industrial renewal is forever a gamble. There can be no guarantee that a renewed metalworking sector will retain the same employment levels or market shares as it has known in the past, only a chance that it will be able to sustain itself at some level in the face of shifting structures of supply and demand.[5]

Accordingly, we should emphasize more strongly public policies aimed at sustaining high levels of economic growth and business formation and expansion—even at the cost of higher rates of business failure and contraction—than public policies aimed at achieving strong and competitive industries through technologically retrofitting or otherwise revitalizing existing firms. Given the substantial gross—and modest net—increase in establishments in the U.S. metalworking sector amid generally threatening business conditions in recent years (see table 3), even the infusion of sophisticated new production capabilities can be expected to proceed at least as rapidly through new business formations as through tax code adjustments aimed at getting older firms to do things differently. This conclusion does not assume that the investment tax credits and accelerated depreciation allowances of 1981 legislation have not triggered a new surge of capital investment throughout the sector; indeed, there is some evidence that precisely that has taken place. A more detailed appreciation of the distinct stages of the extended innovation process, however, shows us that such investment need not lead directly to increased productivity within firms, given the low levels of implementation. Through it all, new business formations have continued, however, and that can be interpreted as a positive sign of achieving the same result in an alternative way. As the current policy debate gives increasing consideration to varieties of protectionism, we must hope it will be realized that policies aimed at preserving an economy in which industries can continue dramatic unorchestrated aggregate adjustments is more important than preserving any particular industrial arrangement. Keeping the goose alive is far more valuable than trying to save a few golden eggs.

Ultimately, the effects of a changing economy on all Americans—those who are labor force participants and those who are not—should be the touchstone of national policies. Strong industries are vital to continued high living standards and the promise of abundant employment opportunities. But the source of industrial strength is flexibility, not rigidity or even stability throughout a sector or industry. It cannot be assumed that policy shifts assigning priority to retaining an economy that can continually adjust through the changing corporate composition of industries must necessarily increase the velocity of changing ties between workers and work. That velocity is already quite high. In the long run, increasingly automated work and workplaces probably pose less threat of net employment loss than the failure to automate. Similarly, it will be the waves of new employment prospects created by new business formations that will allow workers to reattach themselves to a changing industrial structure. With that goal in mind, traditional policy commitments to serving the American worker indirectly by directly assisting existing industries through industrial policies should be reconsidered, just as policies aimed at assisting people indirectly by directly assisting existing settlements through place-oriented urban policies were earlier in this decade. Such indirect approaches are frequently based on obscure links and mighty leaps of faith in explaining how features of an older industrial era, such as existing firms, industrial structures, and urban settlement patterns, must somehow be presumed to preserve best the widest range of opportunities and high living standards.

If the goal of public policy is to serve people first, the emphasis should shift to assisting people—workers and nonworkers alike—more directly. Increasing the options facing Americans as they negotiate the changes accompanying periods of rapid industrial change makes the most sense. I would strongly urge efforts targeted directly to workers and their families to ensure their occupational and geographic mobility. These would include increased emphasis on policies and programs designed to ensure access to social mobility through educational and training and retraining programs as well as to geographic mobility through relocation assistance. The primary aim of public policies should be to identify and increase access by all to private sector opportunities rather than to displace or undercut them by emphasizing the creation of public sector alternatives.

The low incomes and restricted opportunities of many Americans today reflect either their continued unemployment through displacement or their employment in low-wage industries. Generic "mobility" policies need to be devised that can increase these people's prospects of moving up and out of socioeconomic and geographic locations that offer few prospects. Since poverty in this century has increasingly migrated into social locations defined by gender, household structure, and labor skills, as in female-headed households and an underclass of young and unskilled people looking for their first job opportunities, we should make certain that the public policy responses offered include commitments to help people help themselves where they can do so and to provide forms of assistance to those who cannot help themselves.

There is nothing particularly new about industrial change and the adjustments that individuals, firms, entire industries, or even national and regional economies must make to it. The wiser public policy stance is to understand the course and structure of such change and to select public policy responses that bolster the capacities of individual citizens to accommodate it.

Notes

1. National Academy of Engineering, *The Competitive Status of the U.S. Machine Tool Industry* (Washington, D.C.: National Academy Press, 1983), p. 51. See also National Machine Tool Builders Association, *Economic Handbook, 1974–75*.

2. Ibid.

3. See L. Caplan, "Competitive Strategy Not Industrial Policy," *New York Times*, October 17, 1982.

4. L. G. Tornatzky et al., *The Process of Technological Innovation: Reviewing the Literature* (Washington, D.C.: National Science Foundation, 1983), p. 3.

5. See R. S. Greenberger, "Factory-Job Growth Seen in Next 12 Years, but U.S. Calls Outlook Dim in Autos, Steel," *Wall Street Journal*, May 19, 1983.

Appendix
Research Design

The analyses reported in this study derive from data collected between November 1982 and January 1983 as part of a larger study of advanced industrial development and the emergence of computer- and information-based industries in the United States. In-kind support was provided by the University Computing Company (now UCCEL) of Dallas, Texas, and the Numerical Control Society of Glenview, Illinois. This phase of the study was designed to converge and otherwise articulate with a similar study by John Rees and Ronald Briggs, whose consultation and helpful suggestions are gratefully appreciated.[1]

The universe for this study consisted of a portion of the data base of *Modern Machine Shop*, a principal trade publication for the metalworking industry. This publication, established in 1928, is a leading source of information on changes and analyses of their implications for the U.S. metalworking sector. The subscribers it attempts to reach include plants performing metalworking operations as well as those whose products generally emanate from the SIC 25 and 33-39 industry categories. In May 1982 the circulation to individuals and plants in these subsectors exceeded 105,000. The *Modern Machine Shop* master list of machine shops has been compiled from written requests and other communications from subscriber firms; association rosters and directories; business directories; independent field reports; federal, state, and local license records; manufacturers', distributors', and wholesalers' lists; and telephone interviews by the publisher's staff.

A sampling frame for this study was defined so as to include relatively small plants and shops (fewer than 250 employees) that possess an NC or a CNC capability and therefore might face relatively few barriers to expanding their NC/CNC capabilities. Of the 12,523 plants listed in the *Modern Machine Shop* files, 7,994 (63.8 percent) employed fewer than 250 employees and were equipped with at least one numerically controlled machine tool. Survey questionnaires were

TABLE A-1
SURVEY MAILING AND RESPONSE STRUCTURE

Number of plants in universe	12,524
Number of plants in sampling frame	7,994
Number of surveys mailed	7,994
Number of undeliverable surveys	230
Number of surveys returned unusable	83
Number of completed surveys returned	1,172
Response rate for completed surveys	15.3%

sent to each of these plants. Of the surveys returned, the data from 1,172 usable completed questionnaires were coded into seventy-three discrete variables (see tables A-1 and A-2).

TABLE A-2
REGIONAL AND STATE DISTRIBUTIONS OF ALL PLANTS AND SMALL PLANTS IN METALWORKING USING NC/CNC TECHNOLOGY

	All NC/CNC Plants	Small NC/CNC Plants[a]	Small Plant Share (%)
New England			
Maine	36	25	69.4
New Hampshire	108	60	55.6
Vermont	29	14	48.3
Massachusetts	631	390	61.8
Rhode Island	83	37	44.6
Connecticut	401	264	65.8
Middle Atlantic			
New York	916	596	65.1
New Jersey	548	391	71.4
Pennsylvania	948	584	61.6
East North Central			
Ohio	1,180	762	64.6
Indiana	416	241	57.9
Illinois	1,073	711	66.3

	All NC/CNC Plants	Small NC/CNC Plants[a]	Small Plant Share (%)
Michigan	879	573	65.2
Wisconsin	498	287	57.6
West North Central			
Minnesota	322	219	68.0
Iowa	137	80	58.4
Missouri	273	188	68.9
North Dakota	8	4	50.0
South Dakota	17	17	100.0
Nebraska	62	30	48.4
Kansas	121	83	68.6
South Atlantic			
Delaware	12	4	33.3
Maryland	108	67	62.0
Washington, D.C.	1	1	100.0
Virginia	131	67	51.1
West Virginia	43	24	55.8
North Carolina	220	130	59.1
South Carolina	106	55	51.9
Georgia	86	48	55.8
Florida	240	150	62.5
East South Central			
Kentucky	106	63	59.4
Tennessee	121	71	58.7
Alabama	90	47	52.2
Mississippi	42	12	28.6
West South Central			
Arkansas	56	29	51.8
Louisiana	58	33	56.9
Oklahoma	140	89	63.6
Texas	495	330	66.7
Mountain			
Montana	5	1	20.0
Idaho	8	3	37.5
Wyoming	5	2	40.0
Colorado	122	76	62.3
New Mexico	20	19	95.0

(Table continues)

	All NC/CNC Plants	Small NC/CNC Plants[a]	Small Plant Share (%)
Arizona	109	73	67.0
Utah	53	38	71.7
Nevada	13	9	69.2
Pacific			
Alaska	1	1	100.0
Washington	129	87	67.4
Oregon	70	52	74.3
California	1,248	857	68.7
Hawaii	0	0	—
Total	12,524	7,994	63.8

a. Small plants are those with 250 or fewer employees.

SOURCE: *Modern Machine Shop* (Cincinnati, 1982).

Note

1. J. Rees, R. Briggs, and R. Oakey, "The Adoption of New Technology in the American Machinery Industry," *Regional Studies*, vol. 18, no. 6 (1984), pp. 489–504.

Bibliography

Abernathy, W. J. *The Productivity Dilemma*. Baltimore: Johns Hopkins University Press, 1978.

Aldrich, H. E. *Organizations and Environments*. Englewood Cliffs, N.J.: Prentice-Hall, 1979.

Allen, T. J. *Managing the Flow of Technology*. Cambridge, Mass.: MIT Press, 1977.

Armington, C., and M. Odle. "Sources of Job Growth: A New Look at the Small Business Role." *Economic Development Commentary* (Council of Urban Economic Development) (Fall 1982): 3–7.

Austin, D. W., and J. E. Beazley. "Struggling Industries in Nation's Heartland Speed Up Automation." *Wall Street Journal*, April 4, 1983.

Ayres, R. U., and S. M. Miller. *Robotics: Applications and Social Implications*. Cambridge, Mass.: Ballinger Publishing Company, 1983.

Bedell, B. "Aiding Small Business: Give the Money to the Real Innovators." *New York Times*, November 29, 1981.

Bell, D. *The Coming of Post-Industrial Society: A Venture in Social Forecasting*. New York: Basic Books, 1973.

Birch, D. L. *The Job Generation Process*. Cambridge, Mass.: MIT Program on Neighborhood and Regional Change, 1979.

Bluestone, B., and B. Harrison. *The Deindustrialization of America: Plant Closings, Community Abandonment, and the Dismantling of Basic Industry*. New York: Basic Books, 1982.

Boylan, M. G. "The Sources of Technological Innovations." In *Research, Technological Change, and Economic Analysis*, edited by B. Gold. Lexington, Mass.: Lexington Books, 1977.

Braverman, H. *Labor and Monopoly Capital*. New York: Monthly Review Press, 1974.

Bulkeley, W. M. "Computerized Design Systems Being Made for Smaller Firms." *Wall Street Journal*, November 12, 1982.

Bureau of Labor Statistics. *Centennial: 1884–1984, Employment and Earnings*. Washington, D.C.: U.S. Department of Labor, January 1984.

_____. *Productivity Chartbook*. Washington, D.C.: U.S. Department of Labor, 1981.

Business–Higher Education Forum. "The New Manufacturing: America's Race to Automate." Washington, D.C., June 1984.

Bylinsky, G. "The Race to the Automated Factory." *Fortune*, February 21, 1983.

Caplan, L. "Competitive Strategy Not Industrial Policy." *New York Times*, October 17, 1982.

Cohen, R. "The Internationalization of Capital and U.S. Cities." Ph.D. dissertation, New School for Social Research, 1979.

Coleman, J. S., E. Katz, and H. Menzel. *Medical Innovation: A Diffusion Study*. New York: Bobbs-Merrill, 1966.

"Computers in Manufacturing Use, Planned Use, and Buying Influence." *Production* (August–September 1982). Mimeographed.

Corrigan, R. "Beyond the Bottom Line." *National Journal*, June 6, 1983, 1290.

Dostal, W., and K. Kostner. "Changes in Employment with the Use of Numerically Controlled Machine Tools." Mitteilungen aus der Arbeitsmarkt und Berufsforschung, 1982.

Dostal, W., et al. "Flexible Manufacturing Systems and Job Structures." Mitteilungen aus der Arbeitsmarkt und Berufsforschung, 1982.

Downey, H. K., D. Hellriegel, and J. W. Slocum, Jr. "Environmental Uncertainty: The Construct and Its Application." *Administrative Science Quarterly* 20 (1975): 613–29.

Duchesneau, T. D., S. F. Cohn, and J. E. Dutton. *A Study of Innovation in Manufacturing: Determinants, Processes, and Methodological Issues.* Vol. 1. Report to the National Science Foundation. Orono, Me.: University of Maine, 1979.

Duncan, R. B. "Characteristics of Organizational Environments and Perceived Environmental Uncertainty." *Administrative Science Quarterly* (1972): 313–27.

Dun's *Business Month*, February 1984, p. C.

"Emerson Electric: High Profits from Low Tech." *Business Week*, April 4, 1983.

Eveland, J. D. "Issues in Using the Concept of 'Adoption' of Innovations." *Journal of Technology Transfer* 4, no. 1 (1979): 1–4.

Eveland, J. D., E. M. Rogers, and C. M. Klepper. *The Innovation Process in Public Organizations: Some Elements of a Preliminary Model.* Report to the National Science Foundation. Grant No. RDA 75–17952. Ann Arbor, Mich.: University of Michigan, 1977.

"The Factory of the Future," *Newsweek*, September 6, 1982.

Ferguson, E. S. "History and Historiography." In *Yankee Enterprise: The Rise of the American System of Manufactures*, edited by O. Mayr and R. C. Post, 1ff. Washington, D.C.: Smithsonian Institution Press, 1981.

Freeman, C., J. Clark, and L. Soete, *Unemployment and Technical Innovation: A Study of Long Waves and Economic Development*. Westport, Conn.: Greenwood Press, 1982.

Freeman, R. B., and J. L. Medoff. *What Do Unions Do?* New York: Basic Books, 1984.

Gettelman, K. M., and M. D. Albert, eds. "Introduction: Fundamentals of NC/CAM." In *Modern Machine Shop: 1982 NC/CAM Guidebook*, 30–456. Cincinnati, Ohio: Modern Machine Shop, 1982.

Gettelman, K. M., M. D. Albert, and W. Nordquist, eds. "Introduction: Fundamentals of NC/CAM." In *Modern Machine Shop: 1985 NC/CAM Guidebook*, 24–256. Cincinnati, Ohio: Modern Machine Shop, 1985.

Gilpin, R. *Technology, Economic Growth, and International Competitiveness*. Report to the Joint Economic Committee of the U.S. Congress. Washington, D.C., 1975.

Ginzberg, E. "The Mechanization of Work." *Scientific American* 247, no. 3 (September 1982): 39–47.

Gold, B., W. S. Pierce, and G. Rosegger. "Diffusion of Major Technological Innovations in U.S. Iron and Steel Manufacturing." *Journal of Industrial Economics* 18 (1970): 218–41.

Greenberger, R. S. "Factory-Job Growth Seen in Next 12 Years, but U.S. Calls Outlook Dim in Autos, Steel." *Wall Street Journal*, May 19, 1983.

Hannan, M. T., and J. Freeman. "Structural Inertia and Organizational Change." *American Sociological Review* 49 (April 1984): 149–64.

Hanson, R., ed. *Rethinking Urban Policy: Urban Development in an Advanced Economy*. Washington, D.C.: National Academy Press, 1983.

Hicks, D. A. *Advanced Industrial Development: Restructuring, Relocation, and Renewal*. Boston: Oelgeschlager, Gunn & Hain, forthcoming.

Hull, F., and J. Hage. "A Systems Approach to Innovation and Productivity." College Park, Md.: University of Maryland, Center for the Study of Innovation, 1981.

Hunt, H. A., and T. L. Hunt. *Human Resource Implications of Robotics*. Kalamazoo, Mich.: Upjohn Institute for Employment Research, 1983.

International Trade Administration. *A Competitive Assessment of the U.S. Manufacturing Automation Equipment Industries*. Washington,

D.C., June 1984.

Joint Economic Committee. *Robotics and the Economy*. Staff study. Washington, D.C.: U.S. Congress, 1982.

Kamien, M. I., and N. L. Schwartz. *Market Structure and Innovation*. Cambridge: Cambridge University Press, 1982.

_____. "Market Structure and Innovation: A Survey." *Journal of Economic Literature* 13 (1975): 1–37.

Keller, R. T., and W. E. Holland. *Technical Information Flows and Innovation Processes*. Final Report to National Science Foundation. Houston, 1978.

Kennedy, D., C. Craypo, and M. Lehman, eds. *Labor and Technology: Union Response to Changing Environments*. State College, Pa.: Department of Labor Studies, Pennsylvania State University, 1982.

Kimberly, J. R. "Organizational Size and the Structuralist Perspective." *Administrative Science Quarterly* 21 (1976): 571–97.

Lambright, W. H. *Technology Transfer to Cities*. Boulder, Colo.: Westview Press, 1980.

Lawrence, P. R., and J. W. Lorsch. *Organization and Environment*. Cambridge, Mass.: Harvard University Press, 1967.

Lawrence, R. Z. *Can America Compete?* Washington, D.C.: Brookings Institution, 1984.

_____. "Sectoral Shifts and the Size of the Middle Class." *Brookings Review* 3, no. 1 (Fall 1984): 3–11.

Legler, J., and F. Hoy. *Building a Comprehensive Data Base on the Role of Small Business*. Chicago: Heller Small Business Institute, 1982.

Leontief, W. "The Distribution of Work and Income." *Scientific American* 247, no. 3 (September 1982): 188–204.

Mansfield, E. "The Diffusion of Eight Major Industrial Innovations." In *The State of Science and Research: Some New Indicators*, edited by N. E. Terleckyj. Boulder Colo.: Westview Press, 1977.

_____. *Industrial Research and Technological Innovation: An Economic Analysis*. New York: W. W. Norton, 1968.

Mass, N. J., and P. M. Senge. "The Economic Long Wave: Implications for Industrial Recovery." *Economic Development Commentary* (National Council for Urban Economic Development) (Spring 1983): 3–9.

Mayr, O., and R. C. Post, eds. *Yankee Enterprise: The Rise of the American System of Manufactures*. Washington, D.C.: Smithsonian Institution Press, 1981.

McCrackin, B. H. "Why Are Business and Professional Services Growing So Rapidly?" *Economic Review* (Federal Reserve Bank of

Atlanta) (August 1985): 14–28.

Mueller, D. C. "A Life-Cycle Theory of the Firm." *Journal of Industrial Economics* 20 (July 1972): 199–219.

Myers, H. F. "Economists Say Slump Has Hastened Some Trends but Spawned Very Few." *Wall Street Journal*, May 25, 1983.

National Academy of Engineering. *The Competitive Status of the U.S. Auto Industry*. Washington, D.C.: National Academy Press, 1982.

_____. *The Competitive Status of the U.S. Machine Tool Industry*. Washington, D.C.: National Academy Press, 1983.

National Machine Tool Builders Association. *Economic Handbook, 1974–75*.

Nelson, R. R., and S. G. Winter. "In Search of a Useful Theory of Innovation." *Research Policy* (1977): 36–76.

"New Evidence on Small Business Role." *Socioeconomic Newsletter* (Institute for Socioeconomic Studies) 8, no. 5 (August–September 1983).

Noble, D. F. *The Forces of Production: A Social History of Industrial Automation*. New York: Knopf, 1984.

Noyelle, T. J., and T. M. Stanback, Jr. *The Economic Transformation of American Cities*. Totowa, N.J.: Rowman & Allanheld, 1984.

Office of Technology Assessment. *Computerized Manufacturing Automation: Employment, Education, and the Workplace*. Washington, D.C. April 1984.

_____. *U.S. Industrial Competitiveness: A Comparison of Steel, Electronics, and Automobiles*. Washington, D.C., July 1981.

Office of the U.S. Trade Representative. *Annual Report of the President of the United States on the Trade Agreements Program*. Washington, D.C., 1984.

Organization for Economic Cooperation and Development. *Technical Change and Economic Policy*. Paris: OECD, 1980.

Pasmore, W. A., et al. *Sociotechnical Approaches to Organization Change in USAREUR*. Final Report. Cleveland: Case Western Reserve University, 1980.

Peirce, N. R., and C. Steinbach. "Reindustrialization on a Small Scale—but Will the Small Businesses Survive?" *National Journal*, January 17, 1981, 105.

Pelz, D. C., and D. Munson. "The Innovating Process: A Conceptual Framework." Working Paper, Center for Research on Utilization of Scientific Knowledge, University of Michigan, 1980.

Phillips, B. D. "A Guide to Understanding U.S. and State Small Business Job Generation Data." U.S. Small Business Administration,

Office of Advocacy, December 1984.

Premus, R. *Location of High Technology Firms and Regional Economic Development*. Staff study of the Joint Economic Committee. Washington, D.C., 1982.

Rees, J. "Government Policy and Industrial Location in the United States." In *State and Local Finance: Adjustments in a Changing Economy*, vol. 17. Special Study on Economic Change of the Joint Economic Committee. Washington, D.C., 1980.

Rees, J., R. Briggs, and D. Hicks. "New Technology in the American Machinery Industry: Trends and Implications." Washington, D.C.: Joint Economic Committee, U.S. Congress, March 1984.

Rees, J., R. Briggs, and R. Oakey. "The Adoption of New Technology in the American Machinery Industry." *Regional Studies* 18, no. 6 (1984): 489–504.

Riche, R. W., D. E. Hecker, and J. U. Burgan. "High Technology Today and Tomorrow: A Small Piece of the Employment Pie." *Monthly Labor Review* (November 1983): 50–58.

Rogers, E. M., with F. F. Shoemaker. *Communication of Innovations: A Cross-cultural Approach*. New York: Free Press, 1971.

Rosenberg, N. *Perspectives on Technology*. Cambridge: Cambridge University Press, 1976.

_____. "Technological Change in the Machine Tool Industry." *Journal of Economic History* 23 (December 1963): 414–43.

_____. *Technology and American Economic Growth*. New York: Harper and Row, 1972.

Rosenberg, N., and C. R. Frischtak. "Long Waves and Economic Growth: A Critical Appraisal." *American Economics Association Papers and Proceedings* 73, no. 2 (May 1983): 146–51.

Rosenberg, N., and W. E. Steinmuller. "The Economic Implications of the VLSI Revolution." In *Inside the Black Box: Technology and Economics*, edited by N. Rosenberg, 178ff. Cambridge: Cambridge University Press, 1982.

Rybzcynski, W. *Taming the Tiger: The Struggle to Control Technology*. New York: Viking Press, 1983.

Samuelson, R. J. "Business as Usual," *National Journal*, October 23, 1982, 1810.

_____. "A False Religion." *National Journal*, November 20, 1982,1992.

_____. "Middle-Class Media Myth." *National Journal*, December 31, 1983, 2673–78.

Schumpeter, J. *Capitalism, Socialism, and Democracy*. New York: Harper and Brothers, 1942.

Servan-Schreiber, J. J. *The American Challenge*. London: Penguin Press, 1967.

Sharp, M. *The State, the Enterprise, the Individual*. New York: John Wiley and Sons, 1973.

Simon, H. A. *The New Science of Management Decision*. Englewood Cliffs, N.J.: Prentice Hall, 1977.

Small Business Administration. *The State of Small Business: A Report to the President*. Washington, D.C., 1984.

"Small Business Stumbles into the Computer Age." *Business Week*, October 8, 1984.

"Small Is Beautiful Now in Manufacturing." *Business Week*, October 22, 1984.

Smith, A. *The Wealth of Nations*. New York: Modern Library (first edition, 1776), 1937.

Stanback, T. M., Jr., P. J. Bearse, T. J. Noyelle, and R. A. Karasek. *Services: The New Economy*. Totowa, N.J.: Rowman & Allanheld, 1981.

Sternlieb, G., J. W. Hughes, and C. O. Hughes. *Demographic Trends and Economic Reality: Planning and Marketing in the '80s*. New Brunswick, N.J.: Center for Urban Policy Research, Rutgers University, 1982.

Swardson, A. "U.S. Jobless Rate Hits 10.8% in November." *Dallas Morning News*, December 4, 1982.

Terleckyj, N. E. *The State of Science and Research: Some New Indicators*. Boulder, Colo.: Westview Press, 1977.

"The 13th American Machinist Inventory of Metalworking Equipment, 1983." *American Machinist* (November 1983).

Thurow, L. "The Disappearance of the Middle Class." *New York Times*, February 5, 1984.

Toffler, A. *The Third Wave*. New York: Bantam, 1980.

Tornatzky, L. G., J. D. Eveland, M. G. Boylan, W. A. Hetzner, E. C. Johnson, D. Roitman, and J. Schneider. *The Process of Technological Innovation: Reviewing the Literature*. Washington, D.C.: National Science Foundation, May 1983.

Utterback, J. M. "The Dynamics of Product and Process Innovation in Industry." In *Technological Innovation for a Dynamic Economy*, edited by C. Hill and J. Utterback, 40–65. New York: Pergamon, 1979.

Vesper, K. H. *Entrepreneurship and National Policy*. Heller Institute for Small Business Policy Paper, 1983.

von Hipple, E. *The Role of the Initial User in the Industrial Good Innovation Process*. Final report to National Science Foundation. Cambridge, Mass.: Sloan School of Management, Massachusetts Institute of Technology, 1978.

Yin, R. K. "Science and Technology in State and Local Governments: The Federal Role." In National Science Foundation, *The Five Year Outlook: Problems, Opportunities, and Constraints in Science and Technology*, vol. 2, 649–61. Washington, D.C., 1980.

Zaltman, G., R. Duncan, and J. Holbeck. *Innovations and Organizations*. New York: John Wiley and Sons, 1973.

Selected AEI Publications

The Politics of Industrial Policy, Claude E. Barfield and William A. Schambra, eds. (1986, 344 pp., paper $9.95, cloth $20.95)

High-Technology Policies: A Five-Nation Comparison, Richard R. Nelson (1984, 94 pp., paper $4.95, cloth $13.95)

The R&D Tax Credit: Issues in Tax Policy and Industrial Innovation, Kenneth M. Brown, ed. (1984, 47 pp., $4.95)

Trade in Services: A Case for Open Markets, Jonathan David Aronson and Peter F. Cowhey (1984, 46 pp., $3.95)

Protectionism: Trade Policy in Democratic Societies, Jan Tumlir (1985, 72 pp., $5.95)

Essays in Contemporary Economic Problems, 1986: The Impact of the Reagan Program, Phillip Cagan, ed. (1986, about 370 pp., paper $10.95, cloth $20.95)

Futures Markets: Their Economic Role, Anne E. Peck, ed. (1985, 325 pp., $21.95)

Futures Markets: Regulatory Issues, Anne E. Peck, ed. (1985, 376 pp., $24.95)

• *Mail orders for publications to:* AMERICAN ENTERPRISE INSTITUTE, 1150 Seventeenth Street, N.W., Washington, D.C. 20036 • *For postage and handling, add 10 percent of total; minimum charge $2, maximum $10 (no charge on prepaid orders)* • *For information on orders, or to expedite service, call toll free 800-424-2873 (in Washington, D.C., 202-862-5869)* • *Prices subject to change without notice.* • *Payable in U.S. currency through U.S. banks only*

AEI Associates Program

The American Enterprise Institute invites your participation in the competition of ideas through its AEI Associates Program. This program has two objectives: (1) to extend public familiarity with contemporary issues; and (2) to increase research on these issues and disseminate the results to policy makers, the academic community, journalists, and others who help shape public policies. The areas studied by AEI include Economic Policy, Education Policy, Energy Policy, Fiscal Policy, Government Regulation, Health Policy, International Programs, Legal Policy, National Defense Studies, Political and Social Processes, and Religion, Philosophy, and Public Policy. For the $49 annual fee, Associates receive

- a subscription to *Memorandum*, the newsletter on all AEI activities
- the AEI publications catalog and all supplements
- a 30 percent discount on all AEI books
- a 40 percent discount for certain seminars on key issues
- subscriptions to any two of the following publications: *Public Opinion*, a bimonthly magazine exploring trends and implications of public opinion on social and public policy questions; *Regulation*, a bimonthly journal examining all aspects of government regulation of society; and *AEI Economist*, a monthly newsletter analyzing current economic issues and evaluating future trends (or for all three publications, send an additional $12).

Call 202/862-7170 or write: AMERICAN ENTERPRISE INSTITUTE
1150 Seventeenth Street, N.W., Suite 301, Washington, D.C. 20036